Defending the Dinétah

Defending the Dinétah

Pueblitos in the Ancestral Navajo Homeland

RONALD H. TOWNER

THE UNIVERSITY OF UTAH PRESS

Salt Lake City

09 08 07 06 05 04 03
5 4 3 2 1

The Defiance House Man colophon is a registered trademark of The University of Utah Press. It is based upon a four-foot-tall, Ancient Puebloan pictograph (late PIII) near Glen Canyon, Utah.

Library of Congress Cataloging-in-Publication Data

Towner, Ronald H., 1957–
Defending the Dinetah : pueblitos in the ancestral Navajo homeland / Ronald H. Towner.
p. cm.
Includes bibliographical references and index.
ISBN 0-87480-774-3 (alk. paper)
1. Navajo architecture. 2. Navajo Indians—Dwellings. 3. Navajo Indians—History—18th century. 4. Pueblos—New Mexico. 5. Hogans—New Mexico. 6. Dendrochronology—New Mexico. 7. Ethnoarchaeology—New Mexico. 8. New Mexico—Colonization. 9. New Mexico—Antiquities. I. Title.
E99.N3T675 2003
307.3'172'089972—dc21

2003007996

For Byron

Contents

List of Figures

List of Tables

Acknowledgments

First and foremost, I thank Jeffrey S. Dean of the Laboratory of Tree-Ring Research. Virtually all of the ideas in this work, and certainly all of the good ones, were developed talking with Jeff. This project was facilitated in part by the amazing John Roney of the Albuquerque District BLM and James M. Copeland of the Farmington District BLM. I also gratefully acknowledge the assistance of Hugh Rogers, Byron Johnson, Charles Wheeler, Gary Brown, Lori S. Reed, Doug Dykeman, Paul Reed, and Jeff Wharton of Farmington and Steve Fuller, Leslie Sesler, and Tim Hovezak of Dolores. I greatly appreciate the thorough reviews by Steve Nash and Dennis Gilpin. The interest and support of Alan Downer and the Navajo Nation Historic Preservation Department has been very important over the years.

Partial funding for this research has been provided by the National Science Foundation (Grants DBS-9002235, SBR-9002235, and BNS-8504241 to Jeffrey S. Dean, and BCS-9910773 to Jeffrey S. Dean and Ronald H. Towner). The Robert F. Lister Fellowship given by the Crow Canyon Archaeological Center provided crucial funds at a very critical time. Other funding sources include the Laboratory of Tree-Ring Research, the Arizona Archaeological and Historical Society, the Graduate College of the University of Arizona, and the Byron Cummings Award of the Department of Anthropology. Additional monies from the New Mexico Bureau of Land Management and New Mexico SHPO have funded much of the sample dating.

Finally, the interest, skills, and support of Jeff Grathwohl and the staff of the University of Utah Press have been invaluable.

CHAPTER 1

Introduction and Background

> The Navajos consisted of two culturally distinct populations in the first decade or so after 1700—Athapaskans and Puebloans—that were rapidly fusing. Although scholars have tended to view the Navajos as Athapaskans whose culture had absorbed Puebloan cultural traits, we prefer to see them as biological and cultural hybrids, neither Athabaskan nor Puebloan, but a product of both. As a culturally distinct population, the Navajos developed in the Dinétah during the first decades of the eighteenth century (Bailey and Bailey 1986:15).

Bailey and Bailey's assessment of Navajo history and cultural development represents the assumptions, both implicit and explicit, of many archaeologists, anthropologists, and historians for much of the twentieth century. The idea that a large number of Puebloans fled the Spaniards' Reconquest of their Rio Grande villages in the late seventeenth century and helped create Navajo culture has been a standard interpretation for many years. One of the major sources of data used by all three disciplines to support this interpretation has been a class of archaeological sites in northwestern New Mexico known as the pueblitos of Dinétah.

Pueblitos are small stone structures typically built on boulders, mesa rims, and other prominent topographic features (Figure 1). They range in size from one to more than forty rooms, but most contain 4 to 5 rooms and are often associated with forked-pole hogans and a few other structure types. Some sites classified as pueblitos in this study contain no masonry buildings, but consist of forked-pole hogans in inaccessible

Figure 1. The Citadel (LA 55828)

locations protected by small masonry walls or other features. This definition, however, does not include forked-pole hogan sites in open areas that lack associated masonry construction. In addition to the masonry construction, which has been interpreted as an introduction of Puebloan refugees, many pueblitos contain other internal features, such as "hooded" fireplaces, interpreted as non-Navajo introductions. Pueblitos also often contain "loopholes" and "false" entryways that have been cited as evidence of a defensive site function.

Some pueblitos have been investigated since the early days of Southwestern archaeology (Kidder 1913, 1920). Despite having been known in the archaeological literature for nearly eight decades, these sites have never been the subject of a comprehensive archaeological investigation.

This volume is a first attempt to assess a large number of pueblito sites and evaluate their role in Navajo culture using extensive archaeological, dendrochronological, historical, and ethnographic data. As an eighteenth-century phenomenon, and for many years the earliest well-dated Navajo sites, pueblitos have played a critical role both temporally and geographically in interpretations of Navajo culture history and development. Temporally, the tree-ring dates from pueblitos provided confirmation of the early historic period Navajo occupation of the Upper San Juan area described in historical documents. Geographically, the Dinétah pueblitos

Figure 2. Ch'óol'í'í (Gobernador Knob)

demonstrated Navajo occupation of an area from which their territorial expansion could have occurred. The lack of research on pueblito sites themselves, however, has meant that they could be used to support several different theories of Navajo occupation of the northern Southwest and of Navajo cultural development in the Dinétah.

Although masonry pueblitos are not restricted to the Dinétah, the largest number of documented pueblitos (Brugge 1972b; Towner 1996) are located there. The Dinétah is a relatively ill-defined area that includes the Largo and Gobernador drainages of northwestern New Mexico, is known as the Navajo ancestral homeland, and is the location of many of the earliest Navajo archaeological sites. Various definitions of Dinétah have been proposed, including an area within a 30-mile radius of Gobernador Knob (Figure 2), anyplace where Gobernador Polychrome ceramics and pueblitos co-occur, and all lands between the four sacred mountains (Towner and Dean 1996). Following Roessel (1983), in this volume, although the boundaries are still rather ambiguous, I use a relatively conservative definition of Dinétah: the eastern boundary is the continental divide, the western boundary is Blanco Canyon, the southern boundary is approximately the Rio Arriba–Sandoval County line, and the northern boundary is just north of the Colorado–New Mexico state line (Figure 3). The central part of Dinétah includes the Largo, Gobernador, and La Jara

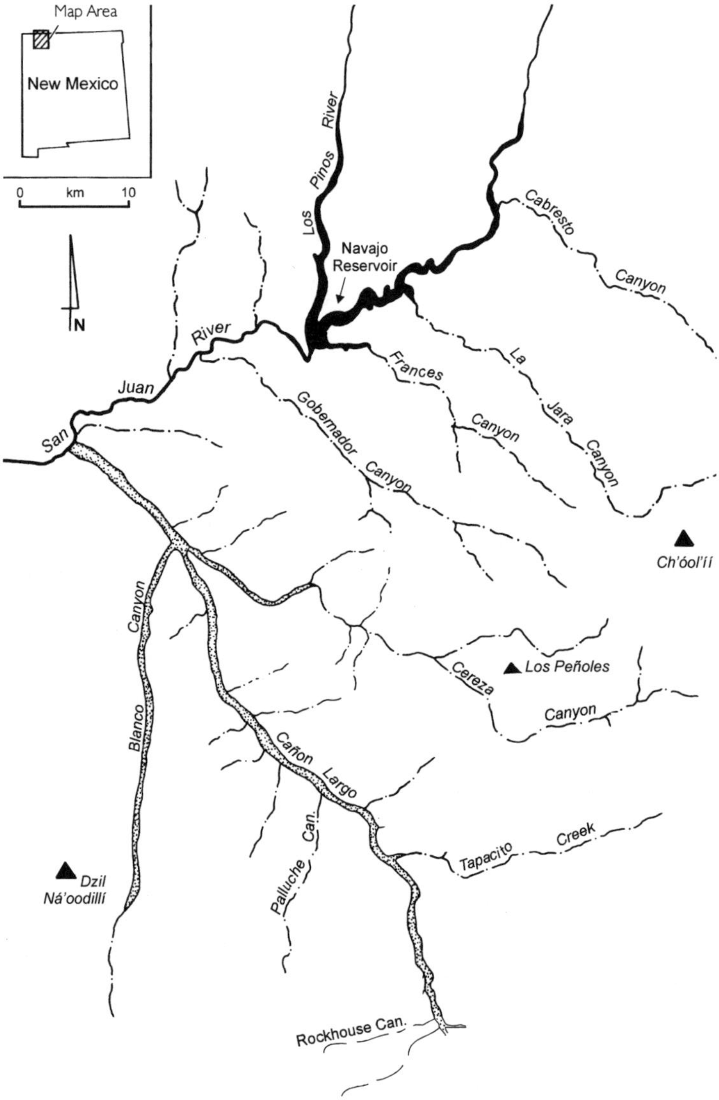

Figure 3. The Dinétah

drainages west of the continental divide; the northern part includes the Navajo Reservoir and drainages, such as Pump Canyon, that enter the San Juan River from the north and extends to the La Plata Mountains in Colorado. This definition purposely excludes areas such as Chacra Mesa (Vivian 1960), Big Bead Mesa (Keur 1941), and Cebolleta (Carroll 1979) that may have supported substantial Navajo populations in the early

1700s. Those areas, however, have not typically been included in definitions of Dinétah (Roessel 1983), primarily because most archaeologists infer that they were occupied later than the Largo-Gobernador area.

PREVIOUS RESEARCH

Throughout much of the twentieth century, early Navajo culture history and adaptation were defined in terms of linguistic models (Hoijer 1938; Sapir 1916, 1936), limited historical documents (Gunnerson 1956; Reeve 1957), and sparse archaeological research. Much of what is known about Navajo culture is derived from historical documents and ethnographies that discuss the period after the Navajos' incarceration at Fort Sumner. In addition, most descriptions of Navajo culture by early ethnographers and anthropologists stress the influences of other cultures on the Navajo. Thus, independent archaeological data from Navajo sites has contributed little to the currently accepted diachronic view of Navajo cultural development. It is only through the use of detailed archaeological, dendrochronological, and other data, however, that a more complete picture of the dynamics of the early Navajo occupation of the northern Southwest can be delineated. The following discussion is necessarily limited, however, to the period from the late 1600s through the early 1800s; this includes the period of occupation of the pueblitos and the abandonment of Dinétah, but omits many changes in Navajo culture that occurred in the nineteenth century. The pueblitos themselves shed little light on the issue of Navajo origins (Schaafsma 2002; Towner 1996; Vierra 1992; Wilcox 1981). As I discuss in Chapter 6, however, other sites in the Dinétah clearly indicate a Navajo occupation of the area in the 1500s, and probably earlier. Although I believe the archaeological and ethnographic data clearly indicate that the Dinétah Navajos entered the area from the north by at least A.D. 1541 (Hancock 1997), and probably prior to A.D. 1450 (Towner 1997; Reed and Horn 1990), the pueblito data themselves contribute little to this issue.

The Navajos in Historical Documents

Early Spanish documents provide glimpses of Navajo culture and history, but because the Navajo lived on the fringes of the Spanish empire, references to the Navajo are sporadic and often ambiguous (Espinosa 1942; Hammond and Rey 1940, 1953; Milich 1966; Winship 1896). The documents must also be interpreted in the context of Spanish objectives and the types of Navajo-Spaniard interaction that occurred. Interaction most often took the form of raids or other military encounters, but there are a few references to Navajo trade with nearby Puebloan groups.

The seventeenth century was one of intermittent conflict between Navajos, Spaniards, Puebloans, Apaches, and various other groups in New Mexico. Navajos and other "non-Christian" Indians were often sold into slavery, both as household servants in New Mexico and as laborers in the mines of Chihuahua (Worcester 1951: 105–06). Slave-raids became particularly common in the mid-1600s; indeed, Governor Bernardo López de Mendizábal claimed an interest in ninety Apache (Navajo?) slaves (Hackett 1937; Scholes 1937). The slave trade earned the enmity of both the Navajo and various Puebloan groups, and there were several attempts at rebellion. The Navajo, however, at different times also raided various pueblos, including Jemez, Santa Clara, Zuni, and possibly Hopi (Bailey 1964; Ferguson and Hart 1985; Forbes 1960; Hodge et al. 1945; Wilson 1985; Worcester 1947). The Spaniards' desire for slaves and retaliation was manifest by several military incursions into the Navajo country (Forbes 1960; Hendricks and Wilson 1996) and probably many other unauthorized expeditions. Ultimately, discontent with the Spanish control of New Mexico culminated in the great Pueblo Revolt of 1680, when the Spaniards were forced to withdraw to El Paso (Kessell 1979). The degree of Navajo involvement in the revolt is unknown, but the power vacuum created by the Spaniards' retreat certainly must have aided the Navajo, both economically and politically.

Traditionally, the Spaniards' reconquest of New Mexico has been viewed as a watershed event in Navajo history. Although don Diego de Vargas's initial reconquest in 1692 had little direct impact, the violent campaigns against the pueblos in 1694 and the crushing of the abortive revolt of 1696 created severe population dislocations in several areas (Hogan 1991). Both historians (Forbes 1960) and archaeologists (Hester 1962a, Kidder 1920) have inferred that the Navajo benefitted from these events. The "refugee hypothesis" (Hogan 1991) suggests that large numbers of Puebloans fled the Spanish Reconquest, particularly after the failed rebellion of 1696, and lived with the Navajo, influencing almost every aspect of Navajo life (Hester 1962a:89).

The return of the Spaniards and the continued raiding by the Navajo prompted various military excursions to the Navajo country, especially after various pueblos had been subjugated. Upon assuming power in

1705, provincial governor Francisco Cuervo y Valdés began preparations for retaliatory expeditions to punish the Navajo for raiding three Tewa pueblos. Roque de Madrid, a seasoned veteran and native New Mexican (Hendricks and Wilson 1996), led the best documented of the campaigns in August 1705. The August campaign entered the Dinétah via Vaqueros Canyon, destroyed Navajo corn fields in La Jara Canyon, and battled with the Navajos at Santos Peak and the confluence of Tapacito Creek and Cañon Largo (Hendricks and Wilson 1996). Roque de Madrid also participated in other Navajo campaigns in 1709 and 1714. Captain Cristóbal de la Serna led campaigns in 1713 and 1716, but little is known about them (Reeve 1958).

After the last campaign, the Navajo and Spaniards were on relatively peaceful terms for most of the 1700s (Reeve 1957, 1959). Beginning about 1716, and possibly as early as 1700, however, conflict with the Ute and Comanche (Hill 1940; Thomas 1935) began to force an accommodation between the Navajo and Spaniards. Conflict with the Ute, Comanche, and possibly other groups continued through the late 1700s and probably played an important role in changing Navajo architectural, settlement, and subsistence patterns (Brugge 1981a, n.d.; Kluckhohn and Leighton 1962); it also may have contributed to the depopulation of the Dinétah homeland by the mid-1750s (Brugge 1983; Towner and Dean 1996). As part of the "Navajo-Spanish Peace" (Reeve 1959), the Navajo expressed an interest in Christianity, and a short-lived mission was built at Cebolleta in 1749 (Hackett 1937). It was abandoned in 1750, however, in part because the Navajo showed little interest in becoming pueblo-dwelling farmers (Reeve 1959:27).

The Dominguez and Escalante expedition of 1776 circumnavigated most of the Navajo-occupied areas (Bolton 1950); the resulting Miera y Pacheco map (Schaafsma 1996: Figure 2.1) shows the Navajo living south of the San Juan river almost as far west as the Hopi villages. Less than a decade later, in 1786, the Navajo were said to consist of five geographic divisions: San Mateo, Cebolleta, Chuska Mountain, Ojo del Oso, and Canyon de Chelly (Bartlett 1932). Ten Navajo groups were recognized by the end of the eighteenth century: Sevolleta, Chacoli, Guadalupe, Cerro Cabezon, Agua Salada, Cerro Chato, Chusca, Tunicha, Chella, and Carrizo (Matson and Schroeder 1957). These changes in the historical references suggest that both the Navajo population and territory were expanding rapidly in the seventeenth and eighteenth centuries.

Significant changes in world affairs in the early nineteenth century impacted the Navajo. Shifting alliances between the Navajo, Ute, Comanche, Puebloans, and Spaniards resulted in alternating periods of conflict and relative tranquility (Thomas 1932). The ten stone towers reportedly built by Navajo leader Antonio El Pinto may have been constructed after the Navajo-Spanish Peace ended in the 1770s (Brugge 1969; Reeve 1960). For many years, these towers were thought to be pueblitos, either in the Dinétah or near Big Bead Mesa, but the research reported here demonstrates that the Dinétah was abandoned prior to A.D. 1760.

A Short History of Navajo Archaeology

Archaeological research in the northern Southwest has been, and continues to be, dominated by questions concerning the prehistoric Anasazi. Research on Navajo archaeological sites has been sporadic throughout this century and has often been driven by nonarchaeological goals. Although this situation has changed somewhat as contract archaeology has expanded, Navajo archaeology is still limited by contract-related concerns and the ability of archaeologists to frame their research questions adequately. Only in the last few years have archaeologists critically evaluated hypotheses generated in other disciplines that relate to questions concerning the Navajo past.

The archaeological knowledge that exists about the Navajo prior to A.D. 1868 has only recently been summarized (Towner and Dean 1996). Prior to 1996, discussions and data concerning Navajo archaeology were uncommon and usually confined to the cultural resource management "gray literature" (Brown 1990, 1993; Brugge 1992). Even the literature that was produced was most often confined to a small area of northwestern New Mexico. The tremendous expansion of interest in and research on Navajo archaeological sites in the last decade has forced archaeologists, anthropologists, and historians to reevaluate their views on pre-1800 Navajo adaptation, cultural development, cultural geography, and interaction with non-Navajos.

Based on surveys and excavations conducted in the Largo-Gobernador area by A. V. Kidder and Nels Nelson, Kidder (1913) presented a paper on Navajo archaeological materials in 1912. But it was not until the 1920s that the first report on Navajo archaeological sites was published (Kidder 1920). Earl Morris excavated at both Old Fort Ruin (LA 1869) and Three Corn Ruin (LA 1871) in the second decade of the twentieth century, but the results were not published until the 1960s (Carlson

1965). The co-occurrence of both stone architecture and forked-pole hogans at the pueblitos indicated to Kidder that "their builders were in contact with the Navajo [and] were members of one of the Pueblo tribes, who for some reason came north" (1920:327). A single historical reference and Jemez oral tradition (Sando 1979a, 1979b) indicated that the cause of the Puebloan immigration was the reconquest of Jemez Pueblo in A.D. 1694 (Bandelier 1890 Part II:215–16, cited in Kidder 1920:328). Kidder's contention that the sites were inhabited by both Navajos and Puebloans became a standard interpretation of Navajo culture history.

R. F. Van Valkenburgh recorded some sites, and Stubbs (1930) and Malcolm (1939) conducted limited surveys of the Gobernador area, but in general, Navajo archaeology languished for more than a decade after Kidder's initial article. Publication of the Rabal document (Hill 1940) and Keur's excavations at Big Bead Mesa (1941) and in the Largo-Gobernador area (1944) marked a resumption of interest in Navajo archaeology. Keur used Kidder's suggestions and her own data to postulate (a) ethnic coresidence by Navajo and Pueblo groups in the Gobernador area and (b) declining Puebloan influence in Navajo culture at Big Bead Mesa. Her data regarding Navajo movements, however, were primarily historical (Twitchell 1911, 1914). She indicates that after a period of initial contact with Puebloan groups, "at Big Bead Mesa the basic Navaho pattern reasserts itself, though probably with enough changes to seriously modify the original culture, and to make a lasting impression" (Keur 1941:71). Although implicit, this is the first example of an archaeologist suggesting migration from the Dinétah to Big Bead Mesa as a process in Navajo cultural development.

Although Malcolm Farmer (1942, 1947) and the Huschers (1942) questioned some of the interpretations put forth by both Kidder and Keur, the basic interpretation of Navajo culture history was established. The publication of tree-ring dates (Douglass 1938; Hall 1951; Stallings 1937) collected from pueblito sites reinforced the idea that those structures were constructed after the Spaniards regained control of New Mexico. Most subsequent work was guided by the ideas that (a) the Navajo entered the Southwest via the High Plains relatively late in prehistory, probably around A.D. 1500 (Hodge 1895; Wilcox 1981; but see Haskell 1987 and Perry 1991), (b) Navajo culture was profoundly influenced by contact with Puebloan groups, particularly after the Spanish Reconquest of A.D. 1694 and failed revolt of 1696 (Forbes 1960; Schaafsma 1981; but see Riley 1954), (c) the Navajo expanded out of the Dinétah and, by implication, west of the Chuska Mountains, only after A.D. 1750 in response to drought and Ute pressure (Hester 1962a, 1962b), and (d) the early Navajo subsistence base and settlement pattern was dramatically different than that observed in the 1800s. It is important to note, however, that all of these ideas are based on historical documents, not archaeological data; the archaeological data have not been independently evaluated, but rather have been used to support inferences derived from historical analyses.

Two major projects in the 1950s and early 1960s generated abundant data with which to address issues of Navajo archaeology. Both projects were inspired by nonarchaeological concerns, however. Navajo Land Claim (NLC) researchers collected data from archaeological sites as far east as Jemez, New Mexico, as far north as Monticello, Utah, as far west as Havasu Canyon, Arizona, and as far south as Reserve, New Mexico. Although the tree-ring data have provided valuable information (Robinson and Towner 1993; Stokes and Smiley 1963, 1964, 1966, 1969), the complete NLC database remains underexploited because of legal restrictions on its dissemination (but see Correll 1979). Important byproducts of the NLC work include *Navajo Pottery and Ethnohistory* (Brugge 1963, 1981b), "The Navajo Exodus" (Brugge 1972b), and Kemrer's (1974) dissertation on western Navajo settlement. Perhaps the most important of these works is *Navajo Pottery and Ethnohistory,* a slender volume that provided the first classificatory system for Navajo ceramics and tied ceramic changes to cultural events and processes. It is still the major reference for Navajo ceramic identification.

The Navajo Reservoir Project (Eddy 1966) brought together a group of specialists who established a Navajo cultural chronology based on archaeological evidence and suggested criteria by which Navajo archaeological material culture could be identified and classified (Dittert et al. 1961; Eddy 1966; Hester 1962a; Schaafsma 1963). Dittert (1958a, 1958b) defined the Dinétah phase, a pre-Revolt Navajo occupation of the Navajo Reservoir area during the project. The Dinétah phase was never adequately dated, however, and was defined by the absence of specific traits associated with the later Gobernador phase that were attributed to Puebloan refugees. Only the presence of corn pollen in sediments thought to date between A.D. 1300 and 1700 (Schoenwetter and Eddy 1964) and the presence of grayware sherds stratigraphically below a single painted sherd at Todosio Rock Shelter (LA 4298) (Hester and Shiner 1963) supported the designation of a pre-Revolt Navajo phase. The Dinétah phase was deleted from the final Navajo Reservoir Project summary (Eddy 1966) because it was poorly dated and possessed neither a type site nor identifiable artifacts with which to compare it to other Navajo cultural manifestations.

Hester's (1962a) landmark publication, *Early Navajo Migrations and Acculturation in the Southwest*, summarized Navajo archaeology and culture history using data from the Navajo Reservoir, Largo-Gobernador, and Big Bead Mesa areas and suggested, based on historical references, that the Navajo entered the Southwest via the High Plains prior to A.D. 1500 (Hester 1962a:73). Hester (1962a:72) rejected the Huschers' (1942) contention that archaeological sites in southwestern Colorado are Navajo, instead preferring to assign them to a Ute occupation. Through the use of both historical and archaeological data, Hester (1962a:67) described the migration of the Navajo out of the Dinétah and presented the most complete Navajo phase sequence available. *Early Navajo Migrations and Acculturation in the Southwest* is still the only publication to directly discuss the Cabezon phase in the Rio Puerco of the East area near Big Bead Mesa.

Shortly thereafter, Carlson (1965) published the results of Earl Morris's early 1900s excavations at several pueblitos. Although often constrained by missing provenience information, Carlson's summary became one of the most important documents concerning the Gobernador phase of Navajo culture history. Carlson was the first to suggest that the pueblitos were constructed primarily by Navajos for defense against Ute raiding. The implications of his arguments, however, were largely ignored.

In the mid-1970s, the National Park Service (NPS) instigated archaeological survey and excavation projects in Canyon de Chelly National Monument, and the resulting publications (Fall et al. 1981; James 1976; Magers 1976; McDonald 1976) documented (a) an early (pre-1800) Navajo occupation of the canyon, (b) changing settlement patterns in response to Spanish and American military pressures, and (c) changing Navajo social and economic systems throughout the pre-Fort Sumner occupation of the area. James (1976) proposed a phase sequence for the de Chelly Navajo separate from sequences developed east of the Chuska Mountains, and Fall et al. (1981) presented a model of changing Navajo settlement and subsistence patterns in and around Canyon de Chelly.

Farther west, Kemrer's research (1974, 1978) used NLC archaeological and tree-ring data to investigate relationships between Navajo settlement, climatic variability, and intergroup interactions on southern Black Mesa. Later research at the Sand Dune Site, near Ganado (Adams 1982; Bradford 1987, Jones 1988), suggested a pre-1750 Navajo presence in the Pueblo Colorado drainage. Recently, Gilpin (1996) summarized Navajo archaeological sites west of the Chuska Mountains; some sites, notably pueblitos, are securely dated in the mid-1700s and it is apparent that "by the 1760s, the Navajo west of the Chuskas were constructing pueblitos, living in large villages, practicing herding and agriculture, and creating rock art identical to that of the Dinétah. Thus, the western Navajo have a longer history in the region than is often recognized and participated in pan-Navajo cultural changes" (Gilpin 1996:171).

Schaafsma (1975, 1978, 1979, 1992, 2002) has proposed the Piedra Lumbre phase near the Abiquiu Reservoir as the earliest Navajo archaeological manifestation in the Southwest. The proposed phase linked the Athapaskans (Querechos) seen by Coronado on the High Plains with the Navajo occupation of the Upper San Juan drainage. Schaafsma (1975; 2002) identified the sites as Navajo based on the presence of (a) subcircular, single room, single-course masonry structures, (b) associated corrals, (c) trash dumps, (d) interior and exterior hearths, and (e) lithic reduction loci. The sites lack Navajo ceramics, however, and are only weakly dated to the seventeenth century. The ceramics present are all Puebloan types (mostly Tewa wares) and may range from the 1600s until 1900, although two types end prior to A.D. 1750. Four noncutting tree-ring dates that range from 1668vv to 1760vv provide little temporal control and were interpreted as intrusive from a later occupation. Schaafsma (2002) eventually refined the Piedra Lumbre phase to extend from A.D. 1640 to 1710. Recently, Binggeli et al. (1998) reported three 1787 cutting dates from a Piedra Lumbre component at LA 13674; such dates suggest the phase may actually be a post-1750 phenomenon. Many archaeologists are reluctant to accept the Piedra Lumbre sites and phase as early Navajo, however. Indeed, based on both historic documents and ceramic analyses, other researchers (Carrillo 1992; Kemrer 1992; Wozniak 1992) interpret the sites as Tewa sheepherding camps.

Marshall's (1985) summary of two Navajo sites in Blanco Canyon, and particularly the radiocarbon dates in the mid-1500s, suddenly brought the Dinétah phase back into existence. Hogan (1989) and Winter and Hogan (1992) used these and other early dates on Navajo sites in the La Plata valley (Hancock 1992; Hancock et al. 1988) to strengthen their arguments for the validity of the phase. Research at the Grassy Canyon site (LA 80854) (Brown et al. 1992) and LA 61852 (Brown 1991, 1996) indicates a pre-Revolt Dinétah phase Navajo occupation of the La Plata Mine area near the Colorado–New Mexico border. Finally, Hancock's (1997) identification of a Navajo forked-pole hogan that dates to the spring of 1541 and Sesler and Hovezak's (1996) documentation of numerous pre-1680 forked-pole hogans north of the San Juan River clearly indicate that the Navajo were present in the Dinétah prior to the Pueblo Revolt. Schaafsma (1996, 2002), however, questions the ethnic affiliation of these sites, preferring to interpret them as protohistoric Ute sites. His dismissal of the sites as Navajo is based primarily on historical documents, not the archaeological data (Schaafsma 2002). If the ethnic identity and dating of these sites indicate an early Navajo occupation near the Colorado border, earlier models of Navajo immigration and adaptation will need to be substantially revised.

Previous Pueblito Research

Pueblitos have played an important role in the interpretation of Navajo culture history for much of the twentieth century. Only recently, however, have comprehensive studies of pueblitos and the areas surrounding them been conducted (Marshall 1991, 1995; Powers and Johnson 1987; Towner 1996, 1997; Towner and Johnson 1998; but see Brugge 1972b).

Kidder's (1920) research on pueblitos and hypothesis that they were evidence of Puebloan and Navajo ethnic coresidence have been standard interpretations for much of this century. Keur's (1941, 1944) work in the Largo-Gobernador and Big Bead Mesa areas linked two of the major pre-1800 Navajo occupations and provided a timeline of events in Navajo history. She suggested that the Athapaskans entered the Southwest via the High Plains just prior to the Spanish Entrada. Her evidence, however, was historical not archaeological. In terms of Navajo history, she viewed the Largo-Gobernador area pueblitos along the same lines as Kidder's hypothesis of ethnic coresidence and, most importantly, suggested that Big Bead Mesa represents a later occupation, implying the movement of people from the former to the latter in the eighteenth century. Hester (1962a:84) indicates that the Navajo population shifted away from the Dinétah toward Big Bead Mesa and Canyon de Chelly about A.D. 1750; there is some suggestion that both areas were partially populated by immigrants from the Largo-Gobernador area.

Hester and Shiner (1963) documented two small pueblitos on the north side of the San Juan River during the Navajo Reservoir Project. At least one of the sites (LA 4314) was occupied in the "late Gobernador Phase" (Hester and Shiner 1963:31) and the other (LA 4331) "was not abandoned until sometime between A.D. 1750–1775" (Hester and Shiner 1963:36). LA 4331 demonstrated multiple masonry styles and a room abutment pattern suggesting multiple construction episodes. They postulate that, because these pueblitos were smaller than those in the Largo-Gobernador area, they either housed fewer people or were seasonally occupied. Unfortunately, neither site yielded tree-ring dates, and both ceramic assemblages were dominated by Navajo-style pottery. Because the area contained only a few pueblitos, and Gobernador Polychrome and Puebloan ceramics were uncommon, Hester and Shiner (1963:73) concluded

> The Navajo Reservoir is at the extreme northern edge of the area inhabited by Pueblo refugees after the revolt. . . . It is apparent that if Pueblo refugees occupied the Navajo Reservoir, they must have been few in number, living as isolated family units. . . . It is clear that the major Navajo-Pueblo acculturation did not occur within the boundaries of the Navajo Reservoir but took place in the localities immediately to the south.

Vivian (1960) expanded the analysis of pueblito sites, including forked-pole hogans, south into the Chacra Mesa area. Through survey, limited excavations, and collection of tree-ring samples, he was able to show that pueblitos in that area dated to the 1730s and later and were probably built for protection against Ute and Comanche raiders (Vivian 1960:218). Most importantly, Vivian (1960:229) suggested that the lack of archaeological evidence for a massive post-Revolt Puebloan immigration indicated that "while Pueblo peoples were easily absorbed into the Navajo tribe, Pueblo traits and traditions were not easily absorbed into Navajo culture." He also suggests that many Puebloan traits in Navajo culture were derived from captives acquired during Navajo raids, both before and after the advent of pueblitos.

Carlson's (1965) publication of Morris's work was the first detailed study of pueblitos in the Largo-Gobernador area. He examined materials from seventeen different sites, including six pueblitos and a "cliff dwelling." It was the pueblito sites, primarily LA 1871 and LA 1869, that led Kidder (1920), Keur (1944), and Hester and Shiner (1963) to support what came to be known as the "refugee hypothesis" (Hogan 1991;

Towner 1996, 1997). Carlson's interpretation of the few available tree-ring dates was that the sites were constructed more than twenty years after the failed Pueblo Revolt of 1696. Importantly, he inferred that they were built not for protection against the Spaniards but for protection against Ute raiding (Carlson 1965:101). Nevertheless, Carlson still suggested that Puebloan refugees built the pueblitos.

Brugge (1972b, 1996, n.d.) discusses the role of pueblitos both inside and outside the Dinétah in changing Navajo settlement patterns. He recognizes that they were probably not adaptive in wars against European-style armies. The earlier publication (1972b), however, had limited distribution and has been used sparingly. Other limited research projects on pueblito sites in the 1970s include Haskell's (1975) excavations and mapping at four sites and Wilson and Warren's (1974) collection of tree-ring samples from Tapacito Ruin. Enloe et al. (1973) surveyed a number of sites in the Largo-Gobernador area and recommended that the area around San Rafael Canyon be designated as the Gobernador Archaeological District. In 1988, the Bureau of Land Management designated San Rafael Canyon a Special Management Area because of its cultural resources. Only recently, however, have the sites in that area been more fully investigated (Towner and Johnson 1996).

Brugge (1986) examined the Chacra Mesa area and described the long-term changes in Navajo settlement in terms of various cultural and noncultural processes. He suggested that some of the earlier Chacra Mesa settlements were occupied by immigrants from the Largo-Gobernador area, although an indigenous Navajo population existed as well. He also indicates that the pueblitos and associated stone-walled hogans may have been true refuges, components of a biseasonal settlement pattern, or both (Brugge 1986:142), but generally housed nuclear- or extended-family groups. More recently, he asserts that the Chaco-area Navajo settlements "played an important role in uniting a dispersed population" that included the San Juan and Mt. Taylor Navajos (Brugge 1996:256).

Publication of *Defensive Sites of Dinétah* (Powers and Johnson 1987) gave a much-needed boost to research on Navajo pueblitos. Powers and Johnson conducted a reconnaissance survey in much of the Largo-Gobernador area and documented more than 100 pueblito sites; their work demonstrated that the number of pueblitos far exceeds previous estimates, although the geographic distribution of pueblitos during the early 1700s remains limited to the Dinétah. They proposed a classification system for pueblito sites consisting of five types: (Type 1) nondefensive, square masonry room-blocks, (Type 2) defensively oriented boulder top masonry room-blocks with rounded corners and controlled access entryways, (Type 3) irregularly shaped or rounded-corner small masonry units situated on lower benches and canyon-bottom boulders, (Type 4) fortified "cliff dwellings," and (Type 5) fortified mesa remnants that lack masonry and consist only of forked-pole hogans. Powers and Johnson (1987: 127–30) summarized the existing tree-ring dates and noted a gap between the A.D. 1694 dates from Tapacito Ruin, a Type 1 site, and all the other pueblitos, a gap they suggested may have been due to a small sample of dated sites (n=17). They suggest that the defensive requirements changed during the 1700s and the classification reflects an increasing need for defense. Because of Powers and Johnson's baseline work, pueblitos can now be viewed as eighteenth-century phenomena that played a central role in the development of larger Navajo communities (Dykeman and Wharton 1996; Jacobson et al. 1992; Marshall 1991, 1995; Towner 1992, 1996; Towner and Johnson 1998).

More recently, Marshall (1991) conducted intensive surveys around ten pueblitos and discovered that in nine cases they were part of larger communities that often included forked-pole and masonry-based hogans, sweatlodges, sheep corrals, ramadas, and extramural hearths and ovens. His analysis of the surface ceramics on these sites indicate that Puebloan wares occur in proportions of 8 percent or less and are derived from a variety of both Eastern and Western Pueblo groups. His ceramic frequencies are similar to those documented for Big Bead Mesa (Keur 1941) and the Navajo Reservoir sites (Hester and Shiner 1963).

Finally, recent research indicates that many pueblitos have lines-of-sight to other pueblitos, and the occupants may have been participants in visual communication networks designed specifically for defensive purposes. Using a Geographic Information Systems (GIS) approach, Jacobson et al. (1992) analyzed potential visual connections between all pueblitos and concluded that pueblitos were built as parts of two different defensive systems designed to counter Ute "hit-and-run" attacks. The first system was a line-of-sight communication network to provide advance warning of trespassers into the Navajo country; the second system consisted of "heavily fortified architecture as an alternative defense in cases where attack was too swift or stealthful" (Jacobson et al. 1992:124). Interestingly, their analysis indicates that pueblitos with lines-of-sight were more complex, with more residential-type and defense-related features, and that pueblitos without lines-of-site were the most defensively positioned and inaccessible. They suggest that (a) pueblitos without lines-of-site were not occupied for long periods of time, but served as protective fortresses, gathering spots, and true "refuges," (b) the occupants of smaller pueblitos with lines-of-site could send warnings throughout the system, and (c) larger sites, even those with fewer lines-of-sight, were used to house Navajos who fled the smaller sites in times of stress (Jacobson et al. 1992:124). Although parts of the analysis must be revised based on more recent tree-ring data, theirs is the most explicit formulation of ideas concerning pueblito functions on both the site and regional levels.

THEMES IN NAVAJO ARCHAEOLOGY

Towner and Dean (1996) identify four basic themes in Navajo archaeology: origins, culture change, expansion, and adaptation. This work examines the last three issues, culture change in the Dinétah, expansion out of that limited area, and early Navajo adaptations to the changing physical and social environments of the eighteenth century. It also has implications for the first issue, Navajo origins.

Navajo Culture Change in the Dinétah

A major issue in Navajo archaeology has been identifying the mechanisms of Navajo culture change. If the "Querechos" described by Coronado were ancestral to the Navajo (Schaafsma 2002), the earliest historical references describe them as Plains bison hunters (see Habicht-Mauche 1992 and Wilcox 1981 for a discussion of this problem). Documents written 200 years later (Hill 1940) describe vast Navajo corn fields in the Largo-Gobernador area; finally, the Navajo of the nineteenth century were noted for their weaving skills and ownership of large sheep and goat herds. Such dramatic changes in subsistence and settlement patterns undoubtedly influenced Navajo culture. Nonarchaeological data, however, have been used to identify the mechanisms of these changes. For much of the twentieth century, scholars used scant historical references to interpret the changes primarily as the result of immigration of other peoples, most notably Puebloan groups fleeing the aftermath of the Spanish Reconquest the 1690s (Bailey and Bailey 1986; Hester 1962a).

In particular, the "refugee hypothesis" (Hogan 1991; Towner 1996), was developed to explain the presence of numerous Puebloan "traits" among the Navajo. Although Bailey and Bailey's (1986) position that the

Navajo did not develop until the first decade of the eighteenth century may be an extreme one, the idea that Athapaskan/Navajo culture was substantially changed by a massive influx of refugees has been, and continues to be, a major explanation for the incorporation of Puebloan traits into Navajo culture (cf. Schaafsma 1996; but see also Schaafsma 2002).

Three aspects of the pueblito sites have been used to support the refugee hypothesis: the supposedly non-Athapaskan masonry architecture (Brugge 1968b), the advent of Gobernador Polychrome ceramics supposedly "Navajo made but Pueblo inspired" (Keur 1941:85) in the early eighteenth century, and the tree-ring dates suggesting post-Revolt pueblito construction (Hannah 1965). The masonry architecture has never been subjected to detailed analysis and comparison with other architectural forms, but is a mixture of Spanish, Puebloan, and Navajo styles; it is very dissimilar to contemporary Puebloan architecture (Brugge 1968b; Hogan 1991; Jett and Spencer 1981; Mindeleff 1898). Hogan (1991:7) suggests that the Navajo copied prehistoric architecture and "had ample opportunity to observe contemporary Puebloan and Spanish defensive architecture as visitors and traders, and from the uniquely instructive perspective of an attacking force."

The beginnings of Gobernador Polychrome production were once thought to coincide with pueblito construction and the influx of refugees (Carlson 1965:57). The type was thought to represent a product of Rio Grande refugees, probably those from Jemez (Kidder 1920). Recent research on both the dating and stylistic attributes, however, indicate the probability that it was a Navajo product that began to be made long before the Pueblo Revolt (Reed and Reed 1992; 1996). Marshall (1985, 1991) suggests that the pottery is a combination of Tewa, Jemez, Hopi, and Navajo styles and is probably the result of Navajo potters imitating Puebloan styles. Reed and Reed (1992) suggest it is an amalgamation of both Eastern and Western Pueblo stylistic attributes and was the continuation of a "yellow ware" tradition furthered by existing trade alliances. Data also indicate that not only was Gobernador Polychrome manufactured prior to the Pueblo Revolt, possibly as early as A.D. 1630, but that the manufacture of Dinétah Gray began earlier and continued to be made until the late-1700s (Dykeman et al. 2002; Reed and Reed 1996; Sesler, Hovezak, and Wilshusen 2001). Thus, interpretations of the ceramic evidence to infer a massive immigration of Puebloans into the Navajo country are, in all probability, erroneous.

Tree-ring data have long played a role in supporting the refugee hypothesis. Hall and Stallings' (1939) 1690–1694 tree-ring dates from Tapacito Ruin, in particular, seemed to confirm an influx of refugees following the reconquest. Even when Carlson suggested that the pueblito sites were built as a response to later Ute pressures, he still believed that an influx of refugees profoundly influenced Navajo culture. He suggested that

> since the culture shown is obviously a mixture of Pueblo and Navajo traits, . . . we must look elsewhere for a slightly earlier occupation by a mixed Pueblo and Navajo group. Such an occupation has been found in the Navajo Reservoir District (Dittert, Hester, and Eddy 1961; Hester and Shiner 1963) on the San Juan proper on the northern border of the Gobernador District (Carlson 1965:98).

It is important to remember, however, that the researchers in the Navajo reservoir concluded that few refugees occupied the Navajo Reservoir area and that most acculturation occurred in the Largo-Gobernador area.

Thus, although neither the Navajo Reservoir area nor the Largo-Gobernador area (nor any other area) contain *archaeological* evidence to support the refugee hypothesis, it has remained the dominant view of early Navajo culture change. Brugge (1996:262–63) suggests that not "all puebloan infusion need be the result of the Reconquest" and that the origins of Navajo clans provide the strongest line of evidence supporting the refugee hypothesis. I examine the latter assertion in detail in Chapter 6. As for Brugge's first contention, I can agree in general that the "Navajo acceptance of Puebloan and Hispanic traits followed circuitous paths" (Brugge 1996:264) and that the source of Puebloan influence "is more likely to be found in the long history of Navajo-Pueblo relations than in the influx of Pueblo refugees in the closing years of the seventeenth century" (Hogan 1991:22). I must reiterate, however, that since the early days of Navajo archaeology (Keur 1944; Kidder 1920) and continuing through the present (Bailey and Bailey 1986; Carlson 1965; Eddy 1966; Hester 1962a; Powers and Johnson 1987; Schaafsma 1996), the Reconquest of 1694 and failed revolt of 1696 have provided the explanatory mechanism for a *massive* Puebloan immigration into the Dinétah. Only Brugge's (1996) and Schaafsma's (2002) recent ideas have suggested a slow "incorporation" of Puebloans into Navajo culture. All previous models of Navajo culture change have implicitly or explicitly described a sudden, massive immigration of displaced Puebloans into the Largo-Gobernador area as a direct result of the Spaniards' reoccupation of the Rio Grande.

Abandonment and Migration out of the Dinétah

Another major issue in Navajo archaeology has been the emigration of the Navajo out of their Dinétah homeland and expansion throughout large portions of Arizona and New Mexico. Historical documents (Reeve 1959, 1960) and archaeological remains (Brugge 1972a, 1986; Keur 1941; Vivian 1960) confirm Navajo movements to the south in the mid-1700s, but little evidence of either kind exists for vast areas of northern Arizona and southern Utah (but see Gilpin 1996). Bailey and Bailey (1986:16) suggest that the Navajos migrating out of the Dinétah disseminated "their hybrid culture to the Athabaskan bands already occupying these areas." There are two implicit assumptions in this statement. First, there must have been existing Athapaskan populations outside the Dinétah in the 1700s; substantial archaeological (cf. Kemrer 1974) and ethnohistorical (Begay and Roberts 1996) data support this assumption. Second, the Dinétah Navajo population must have been a large one, large enough to expand into several areas (Chacoli, Tunicha, Cebolleta, Chusca, Chella, and others) simultaneously and become the dominant cultural force in each of those previously "Athapaskan" centers very quickly. In contrast, Brugge (1986:143) indicates that "the immigrants from the Dinétah bypassed the Chacra in search of more sparsely settled country . . . farther to the south and west."

The abandonment of Dinétah has been typically attributed to a combination of Ute raiding (Towner and Dean 1996) and drought in A.D. 1748 (Powers and Johnson 1987:5; Reeve 1958:20). Indeed, Marshall (1995:203) states "a severe drought which began in about A.D. 1730 had a major impact on the Navajos by A.D. 1748" and "This drought and continued Ute and Comanche predation appear to have caused . . . their abandonment of the Dinétah District homeland." Again, the reliance on historical documents (cf. Reeve 1958), has meant that this hypothesis has not been tested with either paleoenvironmental or archaeological data.

These themes, particularly drought and predation, are discussed throughout this volume. Using the data from the pueblito sites, I discuss primarily the occupation of the Dinétah during the late 1600s and 1700s, and the depopulation of the area by the mid-1750s. These issues, however, are clearly related to other issues, such as Navajo origins and cultural geography.

CHAPTER 2

Laying a Foundation

This chapter outlines the main research domains, questions, and methods used herein to evaluate the role of the pueblitos of Dinétah in eighteenth-century Navajo culture and lays the foundation for discussions of questions of broader anthropological significance that may be addressed using the dendrochronological data.

There are three principal domains in this research: (1) the temporal and spatial aspects of the occupation of pueblito sites and the region, (2) the relationship between past climate and the Navajo occupation and depopulation of the area, and (3) the processes and adaptational responses of early Navajos to changing social conditions of the eighteenth-century Southwest. Each of these broader research domains is related to the others, and the final section of this volume uses archaeological, anthropological, historical, and oral history data to propose a new model of Dinétah Navajo ethnogenesis in temporal and spatial perspective.

TEMPORAL ASPECTS OF THE PUEBLITO SITE OCCUPATIONS

Pueblito sites have been central to discussions of the timing and routes of Athapaskan entry into the Southwest because they were, for many years, the earliest securely dated Navajo archaeological sites. Even today, some researchers (e.g. Schaafsma 1996, 2002; Wilcox 1988) question the validity of the pre-Revolt, prepueblito Dinétah phase and suggest that either (a) pueblitos are still the earliest documented Navajo sites or (b) the pueblitos are the result of Athapaskan and Puebloan immigration into the area after the Spanish Reconquest in the late seventeenth century (Schaafsma 1996:45), or both. Although recent tree-ring research demonstrates that most pueblitos postdate A.D. 1710 (Towner 1992, 1996, 1997), the data presented here expand the number of dated sites both before and after that time. Significantly, these data provide insight into the contemporaneity of different pueblito sites, the contemporaneity of individual structures and rooms at pueblito sites, and the possible relationships between various sites.

Pueblito Site Dendrochronology

The first problem in this research domain is in determining when individual sites were built. Individual sites are evaluated here in terms of the number and quality of the tree-ring dates, the number and sequence of construction episodes, the duration of the site occupation, and the probable time of abandonment and reoccupation (if any) of individual sites.

The relationship between the masonry pueblitos and forked-pole hogans is also an important research issue. There are at least three different possible interpretations of the relationship. First, if the hogans predate the masonry structures, it is possible that the latter were built as true "refuges" to be used only when conflict or the threat of attack necessitated. Second, if the hogans postdate the masonry structures, it is possible that the masonry structures were first used as refuges, and people moved into hogans only after the threat of conflict abated. Finally, if the hogans and pueblitos are contemporaneous they may be (a) evidence of ethnic coresidence (Kidder 1920), (b) evidence of hogan communities using the masonry structures for storage or refuge, or (c) some combination of the above. Not all pueblito sites are the same, however, and much more research is needed before all the possible relationships between forked-pole hogan communities and masonry structures can be identified.

Viewing the aggregate of pueblito sites in order to examine temporal variability in pueblito site construction raises several questions. When were various pueblitos built? How and when did pueblitos grow (if at all)? Are there discrete temporal periods when pueblitos were constructed and, if so, do they correspond with historically documented social events, such as the Pueblo Revolt, the Spanish Reconquest, or Ute raiding? Whether such events corresponded with climatic events, such as drought or increased precipitation, is discussed in Chapter 5.

Estimating Pueblito Site Occupation Spans

Several theoretical models of site occupation duration have been developed for other Southwestern archaeological complexes (Schlanger 1985, 1990). Until quite recently, however, the occupation duration of pueblito sites has received little attention. Originally thought of as simple "refuges" (Keur 1944; Kidder 1920), a general consensus has emerged in the last few years that most pueblitos were "centers" of discrete Navajo communities (Marshall 1991, 1995) and that they served both habitation and defensive purposes (Jacobson et al. 1992; Powers and Johnson 1987). Questions remain, however, concerning the nature of the pueblito site occupations. Were they permanent, year-round habitations, episodically used refuges, seasonal field houses, or some other occupation type? Indeed, implicit in many discussions of pueblitos is the assumption that they were either very short-term refuges used only during times of strife, or were permanent habitations occupied during the entire Gobernador phase of Navajo history (A.D. 1690–1775). Recently, Marshall (1995) and Towner and Johnson (1998) estimated pueblito occupation duration based on midden volume estimates. Marshall (1995:211) suggests that some sites were occupied for decades, but others served as refuges for relatively short periods. Towner and Johnson (1998: 176–79) suggest periodic abandonment and reoccupation of some pueblito sites and shorter-term occupation of others.

Because few, if any, of the sites in this study have accurate midden measurements, I use the tree-ring data, ethnographic analogies, and experimental archaeology to assess the duration of pueblito site occupations. Although I cannot address all the issues surrounding the nature of the occupations, I am able to estimate the number of years many sites were occupied based on construction, remodeling, and repair events. These estimated occupation spans of the sites are critical in assessing the eighteenth-century population dynamics of the Dinétah Navajo and, although not quantified in terms of the population size, indicate different aspects of the eighteenth-century Navajo occupation and suggest that new models of how the Navajo used the area must be constructed.

Estimating Regional Occupation Density

For many years, scholars have assumed that the Dinétah, and the Largo-Gobernador area in particular, contained a large Navajo population, probably the largest eighteenth-century Navajo group in the Southwest (Bailey and Bailey 1986; Brugge 1986; Hester and Shiner 1963). The bases for this assumption appear to have been (a) Spanish references to the Navajo population and the location of the largest concentration of Navajos (Ayer 1916; Hill 1940) and (b) the "spectacular" nature of the archaeological remains in the area. Indeed, there are approximately 130 identified pueblitos sites, and scores of forked-pole hogan sites in the area, a number that far exceeds documented settlements elsewhere. It is the pueblitos, however, that have attracted the attention of both archaeologists and the general public. Forked-pole hogan clusters that lack masonry structures have been identified in areas outside the Largo-Gobernador area, including near Big Bead Mesa (Keur 1941; Towner 1994), near Cebolleta (Carroll 1979), north and east of Navajo Reservoir (Fuller, personal communication), in the Chacra Mesa area (Brugge 1986; Vivian 1960), and near Canyon de Chelly (Fall et al. 1981). Although few of these sites have been precisely tree-ring dated, they must be considered in any discussion of eighteenth-century Navajo settlement and migration patterns.

The appearance of an extremely large population in the Largo-Gobernador area may be more apparent than real, however. The tree-ring data and individual site occupation span estimates provide the first approximation of eighteenth-century Navajo population dynamics, but not necessarily population size, over a broad area of the Dinétah. By combining occupation span data from all of the sites, I am able to address the issue of regional occupation density and suggest a settlement pattern that insures archaeological visibility, but that does not necessarily indicate a large population. The mobility, including seasonal, annual, and periodic movements, of the pueblito builders must be considered in any discussion of the Dinétah occupation and abandonment. This estimation of a more sparse regional occupation has important implications for the reconstruction of Navajo ethnogenesis discussed in Chapter 6.

Spatial Aspects of the Dinétah Pueblito Site Occupations

Pueblitos have generally been characterized as a Dinétah phenomenon (Brugge 1983; Powers and Johnson 1987), although they are also found in other areas (Brugge 1972b; Towner 1996). Previous attempts to evaluate pueblito distributions in the Dinétah (Jacobson et al. 1992; Powers and Johnson 1987) have not been combined with temporal data. Major questions concerning the spatial distribution of pueblitos in the Dinétah include the following: Are there discrete geographic areas where pueblitos were built at specific times? Do pueblitos form individual communities (Marshall 1991, 1995) or are multiple pueblitos parts of larger communities (Towner and Johnson 1998)? Are pueblitos individual defensive sites or parts of larger defensive networks (Jacobson et al. 1992)? Did different defensive systems operate simultaneously or at different times (Towner and Johnson 1998)? These questions must be answered if I am to address the most fundamental questions concerning pueblitos, including why pueblitos were built and by whom.

Climate, the Occupation and Abandonment of the Dinétah

The Navajo expansion out of the Dinétah has generally been viewed as a post-1750 process concordant with the depopulation of the entire area (Brugge 1972b; Towner 1996; Vivian 1960). Emigration out of the Dinétah has typically, if implicitly, been viewed as a sudden "wave" of Navajos abandoning their homeland in the face of the dual pushes of severe drought and Ute raiding (Marshall 1995). Unfortunately, such explanations have not examined the climate before, during, and after the Navajo occupation of the area, have not gone beyond anecdotal references to climate and/or conflict as causal factors, and have not examined the nature of the Navajo occupation. A dendroclimatic reconstruction is used here to examine the nature of precipitation before, during, and after the tree-ring-dated Navajo occupation of the area. Archaeological and dendrochronological data from the pueblitos, the large tree-ring database in particular, are used to examine the nature of the Navajo occupation, the proposed immigration of Puebloans into the area, and the emigration of the pueblito occupants out of the area. In addition to the Dinétah pueblitos, pueblitos and contemporaneous forked-pole hogan village sites in other areas are used to evaluate the nature and duration of the Navajo occupation and depopulation of the area.

Constructing a Model of Navajo Ethnogenesis

In the past, how Navajo culture developed into the form identified in the 1800s has been interpreted as a result of a series of historical events superimposed over a "basic Athapaskan culture" (Keur 1940). Many interpretations of Navajo culture suggest that the "underlying" features of Navajo language and culture—individualism, fear of the dead, high status for women, and flexibility—were covered with a strong "veneer" of other traits after the Athapaskans entered the Southwest. The mechanisms for the acquisition of this "veneer" have almost always involved immigration of people from other societies into Navajo groups and subsequent modifications and reinterpretations of various aspects of the immigrant culture by "native" Navajos.

The problem with this viewpoint, and it is not confined to Navajo archaeology, is that language, material culture, nonmaterial culture, and genetics are not, and never were, a coherent "package" of attributes that co-occur across space and through time. Thus, the cladistic model (Anthony 1990) of Navajo cultural development is fundamentally flawed and cannot explain the development of Navajo culture. In the final section of this volume, I offer a different "ethnogenetic" model (Moore 1994) of Navajo cultural development that I believe better explains the archaeological record and eighteenth century Navajo ethnogenesis.

RESEARCH METHODS: THE ARCHAEOLOGICAL AND DENDROCHRONOLOGICAL DATA

The sites used in this research are divided into five groups: multi-room open-air sites, rockshelter sites, boulder-top pueblitos, mesa rim sites, and defensively oriented forked-pole hogan villages. The data include basic descriptive archaeological data and more detailed dendrochronological data from 65 pueblito sites in northwestern New Mexico (Figure 4). Although efforts have been made to standardize the data, some sites are still not fully recorded and consequently lack some types of data. Maps of the pre-Revolt forked-pole hogans are not included because the exact proveniences of the tree-ring samples are not available. The 65 pueblito sites were selected because they have yielded relevant tree-ring samples and dates and are generally located in the area of the Largo-Gobernador drainages I defined as the Dinétah. Due to space limitations, not all of the pueblito are discussed individually; for additional discussions see Towner (1997).

Archaeological Data: Site Maps and Locations

The majority of sites have been mapped by previous researchers and were not remapped as part of this project. The individual site maps presented herein illustrate specific architectural, dendrochronological, or proveni-

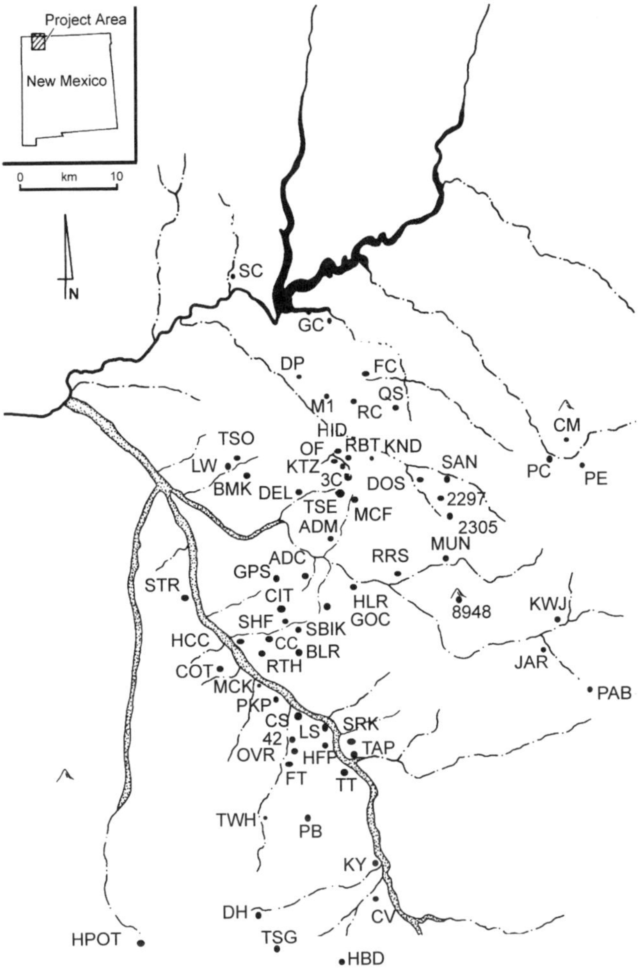

KEY

2297=LA 2297
2305=LA 2305
3C=Three Corn Ruin
42=42 Pueblito
8948=LA 8948
ADC=Adolpho Canyon Ruin
ADM=Adams Canyon Pueblito
BLR=Boulder Fortress
BMK=Buffalo Mask
CIT=The Citadel
CM=Cabresro Mesa Tower Complex
COT=Cottonwood Divide
CRC=Crow Canyon Pueblito
CS=Compressor Station
CV=Canyon View
DEL=Delgadito Pueblito
DH=Deer House
DOS=Dos Cerritos
DP=Dooyiit'íiní Pueblito
FC=Frances Canyon
FT=Foothold Pueblito
GC=Garcia Canyon Pueblito
GOC=Gomez Canyon Pueblito
GPS=Gould Pass Pueblito
HBD=Hummingbird House
HCC=Hadlock's Crow Canyon
HFP=Hooded Fireplace Ruin
HID=Hidden House
HLR=Hill Road Ruin
HPOT=Hanging Pot
JAR=Jarmillo Canyon
KND=Kin Naa dáá'
KTZ=Kin Tsézéí
KWJ=Kin Wólá Jíní
KY=Kin Yazhi
LS=Largo School Pueblito
LW=Lone Warrior Pueblito
M1=Morris 1
MCF=Morris Cliff Dwelling
MCK=McKean Pueblito
MUN=Muñoz Canyon Pueblito
OF=Old Fort Ruin
OVR=Overlook Site
PAB=Pablo Spring
PB=Pointed Butte
PC=Pueblito Canyon Pueblito
PE=Pueblito East
PKP=Pork Chop Pass Site
QS=Quarter-Section Pueblito
RBT=Rabbit House
RC=Romine Canyon Pueblito
RRS=Rincon Rockshelter
RTH=Ridgetop House
SAN=Santo Niño
SBIK=Shándin Bíikin
SC=Simon Canyon
SHF=Shaft House
SRK=Split Rock Ruin
STR=Star Rock Refuge
TAP=Tapacito Ruin
TSE=Tséatah Pueblito
TSG=Tower of the Standing God
TSO=Tso'bito Pueblito
TT=Truby's Tower
TW=Twine House

Figure 4. Locations of the sites used in the study

ence issues; maps of all sites and tree-ring sample proveniences are presented elsewhere (Ababneh et al. 2001; Towner 1997; Towner et al. 2001) and are not reproduced here. Most of the site maps are derived from Powers and Johnson (1987), augmented with specific beam locations; indeed, this research could not have been conducted without Powers and Johnson's baseline work. A few pueblito site maps, created by Jim Copeland and other BLM personnel or the author during the tree-ring collecting, are considered field maps, not precise topographic maps. Other maps, most notably those from Old Fort Ruin (LA 1869), Three Corn Ruin (LA 1871), and Frances Canyon Ruin (LA 2135), are based on Earl Morris's original maps (Carlson 1965), again with individual tree-ring samples added where appropriate; these maps in particular should be viewed as draft field maps. A few sites, generally those sampled only by the Navajo Land Claim, lack maps entirely. *None* of the maps should be considered detailed, precisely scaled representations of the archaeological record, but remapping all 65 pueblito sites was considered beyond the scope of this research. The relationships of individual beams to each other and to the architecture, however, are accurate.

The locations of the sites are derived from individual site forms, New Mexico Archaeological Research Management Section (ARMS) computerized site files, some field checked locations, and published sources. Most are relatively accurate, but some may be in error by a few tens of meters; some of the NLC sites have not been relocated and may be in error by as much as 500 to 1000 meters. Future research, particularly that oriented toward computer or GIS analyses, should not rely on the published site locations, especially for the sites known only from NLC surveys.

Archaeological Data: Architecture

Architecture is an important component of this research at the individual site level. Determining the bond-abut relationships between different masonry walls or architectural units is critical in assessing the presence or absence of anomalous tree-ring dates (Dean 1978). Such an approach is common on Puebloan sites (Dean 1969; Graves 1982; Lightfoot 1992), but is rarely possible on Navajo sites (but see Towner 1997 and Towner and Johnson 1998). Again, because I could not remap all the sites, I assume that the field and published maps are accurate representations of the architecture at the sites. In a few cases, discrepancies were noted in the field between the published map and the archaeological record; these differences are described in the individual site analysis. Because I was particularly interested in the architectural relationships, I gave field observations precedence over published data.

Archaeological Data: Ceramics and other items of Material Culture

The archaeological data concerning items of material culture on the sites are derived exclusively from published sources, augmented by minimal field observations. Although ceramics have played a critical role in studies of the eighteenth-century Navajo sites (Kidder 1920), few quantifiable data are available from the pueblito sites. The most used published source, Powers and Johnson (1987), provides only presence or absence data for various artifact classes, not quantities. Similarly, other sources (Carlson 1965; Keur 1941, 1944) provide no data amenable to statistical analysis. The only data appropriate for such analyses, therefore, are derived from Marshall (1991, 1995). I include the generalized observations simply to show that trade wares comprise a minimal proportion of the ceramics on these sites, despite their use as evidence of ethnic coresidence; other items are mentioned only to indicate possible trade relationships.

Dendrochronological Data

The tree-ring database is composed of samples collected by many different projects and can be divided into two groups: recently collected or reanalyzed samples, and samples collected by early researchers and the Navajo Land Claim that have not been reanalyzed. The majority of the tree-ring samples used were collected by me and Jeffrey S. Dean during the 1990s. Other recently collected or reanalyzed samples were collected as parts of other projects (i.e., Towner and Johnson 1998), in conjunction with other researchers (cf. Dykeman and Wharton 1994), or during special research projects (Ababneh et al. 2001; Towner et al. 2001; Wilson and Warren 1974). These recently collected samples usually include provenience and other data relevant to the proper interpretation of the tree-ring dates.

The second group of tree-ring samples was collected prior to about 1974 and is more difficult to use, primarily because other data, such as provenience, tool marks, and beam attributes, are missing. Tree-ring data of this second type include dates derived by the Navajo Land Claim studies (Robinson and Towner 1993; Stokes and Smiley 1963, 1969), collections made early in the history of dendrochronology (Hall 1951;

Hannah 1965), and other unprovenienced samples (Haskell 1975). The limitations of the second group of samples preclude using them as fully as I might wish. Without the architectural and contextual information, the dates themselves may be subject to misinterpretation and must be used carefully, especially if they are the only samples from a particular site (Towner 1992, 1996). Finally, the Navajo Land Claim samples have not been reanalyzed by the LTRR and must be interpreted differently than the other samples.

Individual dates, beam proveniences, and beam attributes are presented in individual tables for sites or groups of sites discussed in the text. All the remaining samples, typically from poorly documented proveniences, small sites, or single room pueblitos are presented in the Appendix.

Interpreting the Dendrochronological Data

The tree-ring dates, with the exception of some NLC dates, are interpreted using the principles outlined by Bannister (1962), Dean (1978), and Ahlstrom (1985). Although tree-ring dates are accurate and precise to the year, and possibly season, they still must be assessed in their archaeological contexts in order to identify anomalous dates (Graves 1982), define tree procurement and architectural construction events (Dean and Warren 1983; Lightfoot 1992), identify various behaviors associated with wood use (Dean 1996; Towner 1997, 2002), and estimate occupation duration and abandonment times (Ahlstrom 1985; Hantman 1983).

All of the recently collected and reanalyzed tree-ring dates are grouped into the three classes of cutting (or tree death), near-cutting, and noncutting dates (Towner 2002). Cutting dates (symbols B, G, L, r, v) indicate that the last ring on the sample was the last year of tree growth (Robinson et al. 1974). These symbols, alone or in combination, indicate a "true outside" and, in the absence of contrary evidence, the year a tree died. A bridging event (Dean 1978) must be made to infer that the tree died as a result of human activity. Near-cutting dates (symbols +B, +G, +L, +r, +v) indicate that although the sample has a "true outside," one to three rings may be locally absent from the last five years on the sample, but the ring series does not extend far enough past the possibly absent ring to verify its absence. Thus, a "+" date may specify the year of tree death, or that the tree died one to three years after the date given. Like cutting dates, a bridging argument must be made to infer that near-cutting dates represent tree cutting activities. Noncutting dates (symbols vv and ++) do not signify the year of tree death. The symbol "vv" dates indicate that an unknown number of rings may have been removed from the outside of the sample; such rings may have been removed by natural factors such as erosion, or human actions such as shaping or debarking of beams. Thus, tree death occurred some unknown length of time after the vv date. Finally, "++" dates indicate that the ring series cannot be dated beyond a certain point, from which the rings are simply counted rather than crossdated. These "++" dates, even if combined with indicators of true outsides, such as bark, are not crossdated death dates, but ring counts beyond a certain year on the sample and, as such, may seriously underestimate the year of tree death by years or even decades.

The NLC samples, except for the few that have been reanalyzed as part of this project, are accompanied by a different set of symbols. Some symbols, such as "G" (beetle galleries) and "B" (bark present) are the same as those on more recently analyzed samples and indicate the year of tree death. Likewise, an upper-case "C" indicates that the outermost ring is continuous around the circumference of a cross-section and is a cutting date. Many of the NLC samples are designated "+" dates; this + "indicates that either the outermost rings are very small and a ring count only could be made to the outside, or that the outermost dated ring is one just preceding what is usually a small, micro, or absent ring which may be missing from the specimen" (Stokes and Smiley 1963:13). Thus, NLC "+" dates should be considered noncutting dates, although some may actually be near-cutting dates. Many of the NLC dates are accompanied by an "inc" designation, identifying the last ring on the sample as "incomplete." Although this term currently identifies terminal ring attributes that may be related to procurement season (see below), in the NLC case all "inc" designations should be considered noncutting dates, similar to vv dates.

Identifying tree procurement episodes is accomplished by distinguishing trees that were cut while living from those that were harvested as dead wood. Live-cut trees should yield cutting dates, if other factors do not remove outside rings; all dates from samples with true outsides, however, do not necessarily indicate live tree harvesting. Ahlstrom (1985:56–57, 614–17) suggests that, in general, dead-wood samples produce "++" dates, although this is not always the case. Architectural construction episodes are defined as a cluster of cutting dates in the same architectural space; Towner (1992:64) defined the number of dates and type of space necessary to identify a cluster for pueblito sites differently depending on the level of provenience information available. During this project, a strong cluster is defined as three or more cutting dates in the same year from the same architectural space; a weak cluster may contain fewer dates, include noncutting dates, or be only weakly associated with architectural space.

The distinction between a construction episode and a date cluster is important for properly interpreting the pueblito site data. A date cluster, as defined above, depends primarily on the dates and secondarily on the provenience of the samples. As is shown below, identifying a construction episode depends heavily on the provenience of dated samples; there are several cases where date clusters identified previously (Towner 1996: 158–61) can be shown not to represent construction episodes, because of stockpiling, reuse of beams, or other behavioral factors.

Documenting various wood-use behaviors depends on identifying terminal ring characteristics (procurement seasonality), species (tree selection), tool marks (tree or beam modification), and anomalous dates (beam stockpiling and reuse, dead-wood use, and structure repair and remodeling). Such behaviors may also be identified, in part, by the archaeological context of the wooden elements. Terminal ring characteristics are indicated by the symbols "inc" and "comp" on the recently collected or reanalyzed samples, and by "c" on some NLC samples. Incomplete terminal rings (inc) may indicate tree death during the growing season if accompanied by cutting or near-cutting date symbols; if accompanied by noncutting date symbols, the condition of the outer ring is most likely a result of the removal of outer rings by human or natural agencies. Complete terminal rings (comp) indicate tree death after the end of the growing season and, in most cases, are accompanied by cutting or near-cutting date symbols.

Site occupation duration and abandonment are difficult to establish, even given the precise nature of tree-ring dates. Regarding site occupation span, Hantman (1983:143) espouses the formula of subtracting the date that provides the best evidence of initial construction from the latest date and adding a constant of ten years. He does not recognize, however, the value of construction clusters and often uses anomalous dates to calculate occupation span. Hantman (1983:123) also suggests that "when beams are no longer recorded within a site over the course of several years, we can assume that the site was abandoned at or about the date of the latest tree-ring date recorded at the site." Ahlstrom (1985:653–59) points out the limitations of such a "cook-book" approach to interpreting tree-ring dates. He provides several examples indicating that "a gap in the middle of a site's date distribution may indicate a period of abandonment" and should not be included in the computation of occupation spans. Regarding abandonment dates, Ahlstrom (1985:658-59) justifies the use of the latest date as an abandonment date only if (a) the date distribution is relatively continuous and ends abruptly, (b) the latest occupied areas of a site are included in the sample of dates, and (c) the "latest dates from several sites are mutually reinforcing." Many of the pueblito sites in the Dinétah provide excellent examples of such short-term abandonment.

DISCUSSION

The research questions and methods outlined above are designed to delineate significant aspects of the Navajo occupation and abandonment of the Dinétah area. As the earliest positively identified area of Navajo occupation in the Southwest, the Dinétah has played a critical role in interpretations of Navajo culture history and development since early in the twentieth century. The pueblito sites in particular have been a critical part of the database for models of Navajo culture processes used by archaeologists, anthropologists, and historians. This research, however, is the first detailed analysis of the data from a large number of pueblito sites. By examining the temporal and spatial aspects of the Navajo occupation of pueblitos during their construction and occupation, and addressing possible causes for their abandonment, I provide an expanded database from which to assess seventeenth- and eighteenth-century Navajo cultural dynamics across a broad area of the Southwest.

The changes in our interpretations of Navajo cultural development proposed as a result of this analysis are important for elucidating not only the Navajo occupation of the area, but for examining the protohistoric period throughout the northern Southwest. In addition to their relevance to Navajo cultural development and adaptation, I believe the pueblito site data can be used to provide information relevant to a broad range of methodological and theoretical issues important in anthropology and archaeology. Broader theoretical issues include understanding culture change, intra- and intergroup relationships, demographic trends, processes such as emigration and immigration, the interaction of people with their environment, and many other aspects of human existence. The pueblito data may also be used to develop models that allow us to better identify some of these processes in prehistoric contexts where historic sources are lacking and ethnographic analogy far more tenuous (Brugge 1996).

CHAPTER 3

The Dinétah Pueblito Sites

> The stone towers of the Dinétah have long held a fascination for archaeologists, and almost a century of research has been devoted to understanding the cultural processes that gave rise to these monuments (Hogan 1991:i).

Despite Hogan's contention, the pueblitos themselves have not been the subject of intensive archaeological research, but have been used to support hypotheses based on other types of evidence. Pueblito sites are presented here in five different groups: defensively oriented forked-pole hogan sites, large, well-dated pueblitos, boulder-top pueblitos, rockshelter sites, and mesa rim sites. Other than the rockshelter and defensively oriented hogan sites, the groups are somewhat arbitrary. Many mesa rim sites are also on top of boulders, and some of those sites are well dated. Nevertheless, these groups make some sense, and I leave it to someone else to construct a more rigid pueblito classificatory system if they so choose.

The pueblito sites used in this analysis include many of those masonry structures and associated forked-pole hogans in the Dinétah that have yielded tree-ring dates and comprise the largest number of such structures in the Southwest; 65 pueblito sites are described in Towner (1997), but are not described here because of space limitations. Although there are still sites yet to be sampled in the Dinétah, these data include absolutely dated tree-ring samples from approximately 50 percent of the known pueblito sites; such a sample size must surely be deemed adequate to address the research questions outlined in Chapter 2.

Figure 5. A forked-pole hogan on Frances Mesa

DEFENSIVELY ORIENTED FORKED-POLE HOGAN SITES

One group of sites included under the rubric of pueblito sites is better described as defensively oriented forked-pole hogan sites. These sites—Cottonwood Divide, Star Rock Refuge, and Hanging Pot—are all located on rocky prominences or isolated ridge lines with expansive viewsheds and very limited access routes. They consist primarily of forked-pole hogans (Figure 5), and often include small masonry wall segments near the perimeter of the rock, crevices covered with rocks and stones, and, occasionally, small masonry storage rooms. All these sites are located west of the Largo drainage (Figure 4). None of these sites are amenable to architectural analysis because all structures are free-standing hogans. The tree-ring data from the defensively oriented hogan sites are presented in Table 1. Only Hanging Pot is discussed here as an example of the site type.

Hanging Pot (LA 2445)

Hanging Pot is a scattering of hogan wood on a small butte overlooking the badlands of upper Betonnie Tsosie Wash and another hogan at the base of the eastern side of the butte (Figure 4). The site has gained some

Table 1. Tree-ring Samples Collected from Defensively Oriented Forked-pole Hogan Sites

Site Name Number	Sample Number	Species	Provenience	Inside Date	Outside Date	Terminal Ring	Tool Marks	Comment
COTTONWOOD	DNT-395	PNN	Hogan leaner 2	1599	1712vv	inc	Indet	weathered
DIVIDE	DNT-396	PNN	Hogan leaner 2	1619	1715vv	inc	Indet	weathered
LA 55829	DNT-397	PNN	Hogan leaner 2	NO DATE	—		Indet	weathered
	DNT-398	PNN	Hogan leaner 2	1618p	1707+vv	inc	Indet	weathered
	DNT-399	PNN	Hogan leaner 1	1387p	1633++vv	inc	Indet	weathered
	DNT-400	JUN	Hogan leaner 1	NO DATE	—		Indet	weathered
	DNT-401	PNN	Hogan leaner 3	1414	1637vv	inc	MA end	weathered
	DNT-402	PNN	Hogan 4	1369	1624vv	inc	Indet	weathered
	DNT-403	JUN	Hogan 4	NO DATE	—		Indet	weathered
	DNT-404	PNN	Hogan 4	1622	1715vv	inc	Indet	weathered
HANGING	DNT-671a&b	JUN	Loose log Area 2	NO DATE	—		MA ends	fork
POT	DNT-672	PNN	Loose log Area 2	1612	1740vv	inc	MA end	fork
LA 2445	DNT-673	PNN	Loose log Area 2	1542	1747vv	inc	MA end	
	DNT-674	JUN	Loose log Area 1	NO DATE	—		MA end	
	DNT-675	PNN	Hogan fork 1	NO DATE	—			spiral; root
	DNT-676	PNN	Hogan fork 1	1617	1735++vv	inc	MA end	
	GP 5979	PP	Unknown	1302np	1480++vv	inc	Unknown	GP date 1783v
STAR	DNT-599	JUN	Loose log	NO DATE	—		MA ends	
ROCK	DNT-600	PNN	Loose log	1392	1708vv	inc	MA end, broken end	
LA 55838	DNT-601	JUN	Door lintel	NO DATE	—		MA end	split
	DNT-602	JUN	Door lintel	NO DATE	—		Broken end	
	DNT-603	JUN	Door lintel	NO DATE	—		MA end	
	DNT-604	JUN	Door lintel	NO DATE	—			
	DNT-605	JUN	Core destroyed	NO DATE	—			
	DNT-606	JUN	Loose log (upright)	NO DATE	—		MA limb, shape, end	flat
	DNT-607	JUN	Horizontal intramural	NO DATE	—		Broken end	
	DNT-608	JUN	Loose log	NO DATE	—		MA end	
	DNT-609	JUN	Loose log (upright)	NO DATE	—			

Key: DNT= Tree-ring Lab Collection; GP= Gila Pueblo Collection; JUN= Juniper; PNN= Pinyon; PP= Ponderosa pine; inc= incomplete; comp= complete; MA limb= Metal ax limb trim; MA end= Metal ax-cut end; Indet= Indeterminate; MA bark= Metal ax debarking; MA shape= Mteal ax shaping notoriety as the "Tower of Haskhek'izh" (Spotted Warrior) and the location of a battle where Utes lay siege to a "Navajo watchtower" (Van Valkenburgh 1940), but that interpretation has been questioned (Towner and Johnson 1998). No known pueblito sites are visible from Hanging Pot.

Material Culture. Van Valkenburgh (1940) states that ceramics on the site included Acoma trade wares, but currently only Dinétah Gray wares exist, and no trade ware pottery is present. There is little refuse either within the site boundaries or over the edges of the mesa. Red Woman told Van Valkenburgh (1940) that the site could be identified by three petroglyphs carved under a ledge in the tower site: a jagged line, a St. Andrew's cross, and a jagged line with expanding ends, which are symbols for Male Lightning, a star, and the Ant People, respectively. The petroglyphs have not been located, however, despite an examination of the cliff face below the site (H. Rogers, personal communication).

Tree-Ring Samples and Dates. Seven tree-ring samples have been collected from the site, but none produced cutting dates. Four of the samples (DNT-671–674) are from the area on top of the mesa and two (DNT-675–676) are from a forked-pole hogan near the base of the mesa. The sample collected by Van Valkenburgh (GP-5979) was not provenienced beyond the site level; it was dated by Gila Pueblo at 1630–1783v, but has since been corrected by the Laboratory of Tree-Ring Research and dates 1302np–1480++vv. Two juniper and four pinyon samples were collected during this project; the Gila Pueblo sample is ponderosa pine. All of the documented samples were procured using a metal ax.

Interpretations. Three noncutting dates place the site occupation in the 1740s or later for both the area on top of the mesa and a hogan at the base. The site exhibits no evidence of having housed a substantial number of people or of ever having been repaired. When compared to other pueblitos in the Dinétah, LA 2445 seems to be a site occupied for a short term by very few people.

OPEN-AIR MULTI-ROOM PUEBLITO SITES

Three sites are classified as multi-room open sites—Tapacito Ruin, Old Fort Ruin, and Ridgetop House (Figure 4). Only Tapacito Ruin and Old Fort Ruin are discussed here. As discussed below, Tapacito Ruin is unique and could be classified separately. Old Fort Ruin and Ridgetop House consist of central room-blocks and plaza areas surrounded by enclosing walls. Both sites are located near sandstone bluff edges and could be classified as mesa rim sites; their internal organization, however, distinguishes them from other pueblitos. Cleveland and Kurley-Begay's (in press) research on oral traditions indicates that Old Fort functioned as a refuge for the elderly and injured; it would be interesting to know if Ridgetop House is similarly discussed in other oral traditions.

Tapacito Ruin (LA 2298)

Tapacito Ruin is located on a bench above Canyon Largo near the confluence of Tapacito Creek and Canyon Largo. Jacobson et al. (1992) indicate lines-of-sight to Split Rock Ruin and Truby's Tower, but neither site can be seen. In addition to a seven-room masonry structure, the site complex includes five forked-pole hogans, one masonry-based cribbed-log hogan ring, a rectangular masonry-based structure, three hearth areas, four middens, and a possible pit structure (Marshall 1991). Because of its unique place in the history of pueblito research (Towner 1996), it has played an important, though probably unwarranted, role in the interpretation of the Navajo occupation of the area. It is architecturally, ceramically, chronologically, and topographically different from other pueblitos, and it should be viewed as a separate entity.

Material Culture. Powers and Johnson (1987) noted Dinétah Gray, Gobernador Polychrome, and Zia-Acoma sherds on the site. In addition, Marshall (1995:65–69) noted Rio Grande Glazewares from Puname, Acoma, Cochiti, and perhaps the Galisteo Basin, as well as flaked and ground stone artifacts, faunal remains of domesticated and nondomesticated animals, and European artifacts made of iron, glass, and copper. Interestingly, Tapacito is one of the few pueblitos where Ocate Micaceous ceramics of probable Plains Apache manufacture have been found. The site has one of the highest frequencies of nonlocal ceramics of any pueblito analyzed (Marshall 1991).

Architecture. The site consists of a core of four rooms with standing walls and three areas of rubble that may represent additional rooms (Figure 6). The standing walls are constructed of massive rubble core-veneer masonry approximately 1 m thick and 2.5 m high (Figure 7). Single-

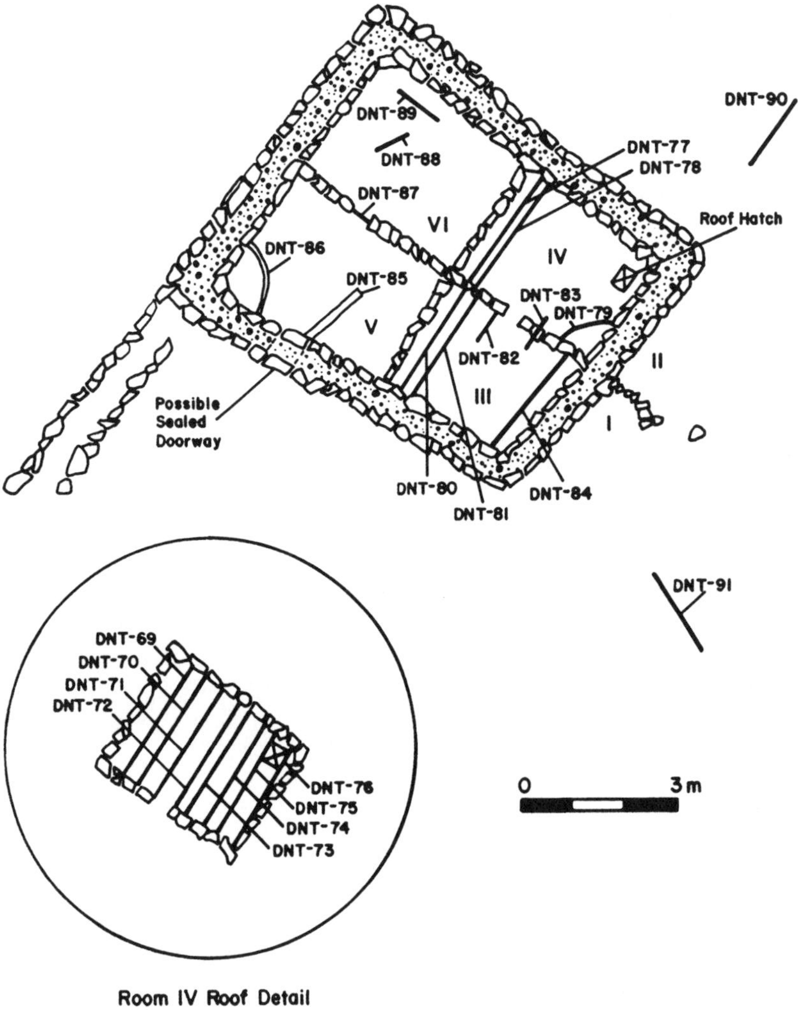

Figure 6. Tree-ring sample proveniences at Tapacito Ruin (LA 2298). (Map from Towner and Dean, 1992)

coursed sandstone walls on top of some of the rooms possibly represent a second story or a parapet. Formerly, portions of the second story stood 60 to 75 cm (2–2.5 feet) high (Brugge and Correll 1959), but are currently reduced to a few stones. Spanish-style hooded fireplaces are present in Rooms IV and V. Small doorways connect Rooms III and IV and V and VI, but direct movement between Rooms IV and VI and Rooms III and V is impossible. There are no outside entrances, and access to the interior must have been through the roof. A roof hatch is still present in Room IV.

Figure 7. Walls at Tapacito Ruin

Tree-Ring Samples and Dates. Three sets of tree-ring samples have been collected from Tapacito Ruin over the past 50 years (Table 2). The first set was collected by Hall and Stallings in 1941, the second by Wilson and Warren (1974), and the third by Towner and Dean (1992). Twenty-nine dates are available for LA 2298. All of the wooden elements sampled by Towner and Dean (1992) show unmistakable evidence of the use of metal axes for tree felling, length reduction, limb trimming, or debarking. All of the beams, moreover, were cut to fit the square rooms of the pueblito and socketed into its walls.

Ten A.D. 1694 cutting dates and six near-cutting dates between A.D. 1691 and 1693 have been documented for the masonry structure. Eight A.D. 1690 cutting dates and one A.D. 1689 cutting date come from pinyon samples collected by Hall and Stallings. Despite extensive resampling of the site, these pinyon specimens have not been duplicated and their origin remains somewhat problematic. The tight clustering and species consistency indicate that they probably were taken from a single structure rather than several different proveniences. Towner and Dean (1992) suggest that the A.D. 1690 samples were collected from one of several hogans documented at the site complex (Marshall 1991).

Table 2. Tree-ring Samples Collected from Tapacito Ruin (LA 2298)

Sample Number	Species	Provenience	Inside Date	Outside Date	Terminal Ring	Tool Marks	Comment
DNT-69	PNN	RM IV Roof primary	1593	1694B	comp	MA limb, end	SB
DNT-70	PP	RM IV Roof primary	NO DATE	—		MA bark, limb	SB
DNT-71	PP	RM IV Roof primary	1563p	1694B	comp	MA bark, limb	SB
DNT-72	PP	RM IV Roof primary	1559	1694LB	comp	MA bark, limb, shape	
DNT-73	PP	RM IV Roof primary	NO DATE	—		MA bark, limb	SB
DNT-74	PP	RM IV Roof primary	NO DATE	—		MA bark, limb	SB
DNT-75	DF	RM IV Roof primary	1311	1561++vv		MA bark, limb, shape	
DNT-76	PNN	RM IV Roof primary	1625p	1692+B	comp	MA bark, limb	SB
GOV-44	DF	RM IV Beam	—	—			same as RG-803/804, DNT-80, 85, 88, GOV-45
DNT-77	PNN	RM IV Shelf pole	1635	1692+B	comp	MA limb	SB
DNT-78	PNN	RM IV Shelf pole	1619p	1689+vv		MA bark, limb	SB
DNT-79	POP	RM IV Roof primary	NO DATE	—		MA limb	SB
GOV-46	PNN	RM I/II Loose log	1401p	1693R	inc		same as RG-815
DNT-80	DF	RM III Roof primary	1550	1694cL	comp	MA limb	same as DNT-85, 88, GOV-44, 45, RG-803, 804
DNT-81	PNN	RM III Roof primary	1627	1693+v	inc	MA limb	
DNT-82	PP	RM III Roof primary	1617p	1694v	inc	MA limb	
DNT-83	PP	RM III Roof primary	16127p	1694B	inc	MA limb	SB
DNT-84	PP	RM III Roof primary	NO DATE	—		MA bark, limb, end	SB
GOV-47	PP	RM III Loose log	1559+p	1694rL	inc	Unknown	
GOV-48	PP	RM III Loose log	—	—	inc	same as DNT-91	
DNT-85	DF	RM V Roof primary	—	—	comp	MA limb, shape, end	flat; same as GOV-44, 45, DNT-80, 88, RG-803, 804
DNT-86	POP	RM V Fireplace	NO DATE	—		MA limb	SB
GOV-45	DF	RM V Beam	—	—	comp		same as RG-803, 804, DNT-80, 85, 88, GOV-44
GOV-43	PNN	RM V Loose log	1565+p	1689vv		Unknown	
DNT-87	PNN	RM V/VI Door lintel	1490	1657++vv		MA limb, shape, end	flat

DNT-88	DF	RM VI Loose log	—	—	comp		same as DNT-80, 85, GOV-44, 45, RG-803, 804
DNT-89	PNN	RM VI Loose log	1526	1683+vv		MA limb, shape	flat
DNT-90	PNN	Exterior Loose log	1396+	1670+vv		MA limb	
DNT-91	PP	Exterior Loose log	1559+p	1694r	inc	MA limb	same as GOV-48
RG-802	PNN	Unknown	1581p	1690cL	inc	Unknown	
RG-803/804	DF	Unknown	—	—	comp	Unknown	
RG-805	DF	Unknown	1455p	1694cL	comp	Unknown	
RG-806	DF	Unknown	1625p	1694rL	comp	Unknown	
RG-807	PNN	Unknown	1603p	1690c	inc	Unknown	
RG-808	PNN	Unknown	1619p	1690c	inc	Unknown	
RG-809	PNN	Unknown	1598p	1690v	inc	Unknown	
RG-810	PNN	Unknown	1628p	1690cL	inc	Unknown	
RG-811	PNN	Unknown	1614p	1690r	inc	Unknown	
RG-812	PNN	Unknown	1659p	1690r	inc	Unknown	
RG-813	PNN	Unknown	1643p	1690r	inc	Unknown	
RG-814	PNN	Unknown	1634p	1689v	inc	Unknown	
RG-815	PNN	Unknown	—	—	inc	Unknown	same as GOV-46
RG-816	PP	Unknown	1637p	1694rL	comp	Unknown	
RG-817	PNN	Unknown	1466p	1691+v	comp	Unknown	

Key:

DNT= Tree-ring Lab Collection
GOV= Douglass Collection
RG= Rio Grande Collection

PP= Ponderosa pine
DF= Douglas-fir
POP= Cottonwood
PNN= Pinyon

MA limb= Metal ax limb trim
MA end= Metal ax-cut end
MA bark= Metal ax debarking
MA shape= Metal ax shaping

SB= smoke blackening present
inc= incomplete
comp= complete

Interpretations. Based on their well-provenienced samples, Towner and Dean (1992) assert that the core of the pueblito was built in a single construction episode in the late summer or early fall of A.D. 1694. Two, or possibly three, peripheral rooms were added after the initial construction event, but were apparently never roofed. The cluster of A.D. 1690 dates from Hall and Stallings's collection does not belong to the pueblito and probably comes from a hogan in the site complex that was built in the summer of A.D. 1690. Marshall (1995:60), citing Towner (1992), rejects this interpretation and states, "this is unlikely given the poor condition of the structures and the probability that Hall and Stallings would have selected the much better beam material from the nearby pueblito." Towner and Dean (1992), however, explicitly address this possibility, as well as two others, and demonstrate that the beams could not have come from the pueblito. Certainly, Hall and Stallings collected "better beam material," but Marshall appears to have ignored that they collected it more than fifty years prior to his assessment of the site and that by taking only beam cross sections they could *not* have sampled the beams in the masonry structure.

Terminal ring analysis reveals that all the 1690 trees from the hogan were felled during the summer pinyon growing season, probably in June or July. In contrast, complete terminal rings show that all the 1694 Douglas-firs from the masonry pueblito were felled after their growing season ended in May or April. A mixture of complete and incomplete terminal rings (three samples each) indicates that the 1694 ponderosa pine logs were cut toward the end of that species' growing season in the late summer when some of the trees had ceased growing but others had not. Thus, the logs used in the masonry pueblito probably were felled as a group early in the autumn of 1694, and the four-room core of the structure was completed before winter.

Several aspects of wood-use behavior at Tapacito Ruin merit attention. First, although the site is located in the pinyon-juniper forest, not a single juniper specimen was used in the construction of the pueblito. Apparently the wood required to roof the masonry structure had to meet certain length and straightness constraints. Pinyon, ponderosa, and Douglas-fir apparently met these requirements, but juniper did not. In addition, a single Douglas-fir tree sampled by Wilson and Warren (1974) (GOV-44, GOV-45) and by Towner and Dean (1992) (DNT-80, DNT-85, DNT-88) was used for roof beams that span Rooms III, V, and VI; thus, the original Douglas-fir log was at least 30 feet long when it was cut and probably required substantial group cooperation to procure and fit into the structure.

A second important aspect of the distribution of species in the pueblito is the location of the cottonwood (*Populus* sp.) samples. The only two cottonwood samples in the structure are from the hooded fireplaces and were probably selected because cottonwood is flexible enough to be bent into curved bases for hooded fireplaces.

Marshall (1995:211) estimates that the site was occupied for approximately twelve years. Given the meager evidence for remodeling of the pueblito and the presence of a hogan prior to pueblito construction, this estimate seems reasonable, although it may overestimate the occupation by a few years.

Finally, the question of the site's role in the Roque de Madrid campaign must be addressed. In tracing the Spaniard's route, Hendricks and Wilson (1996) suggest that the last battle of the campaign occurred at the confluence of Tapacito Creek and the Largo. The Spaniards remained on the valley bottom, but sent some of their Indian auxiliaries up Tapacito Creek to attack the Navajos from behind. Because Roque de Madrid's Indian auxiliaries "could hardly have missed seeing such a structure and undoubtedly would have told the Spaniards about it," Hendricks and Wilson (1996:86) suggest that "the pueblito at LA 2298 had not been constructed in 1705, and whenever it was built, the wood in it was salvaged and used from another dwelling. The intact roof over one room also argues that the structure is more recent in age."

As an archaeologist and dendrochronologist, I am much more inclined to believe the tree-ring dates from the structure than assume it was not built because the Indian auxiliaries did not mention it to the Spaniards. Tapacito Ruin is one of the most securely dated structures in the Dinétah. To propose that any structure was built with (at least) 10 eleven-year-old beams and *not a single freshly cut timber* ignores all of the data collected by dendrochronologists over the past half century and certainly ignores anything known about Navajo wood-use behavior (Dean and Russell 1978). The wood in the structure was cut to fit this pueblito and shows no evidence of having been used in another structure. Simply because one room has an intact roof does not indicate it "is more recent in age." There are many similar structures across the Southwest that are nine hundred years old, yet still have intact roofs.

There are several ways to explain both the battle as described and the pueblito as present when the battle occurred. First, Tapacito Ruin is not precisely at the confluence of Tapacito Creek and the Largo, but approximately three-fourths of a mile north of the confluence; it is also more than "about 20 meters back from the [bluff] edge" (Hendricks and Wilson 1996:85). In addition, although the area around the site is usually described as a bench, it is better described as a structural bench covered with undulating aeolian hills. Indeed, this single story masonry structure, not located on a boulder or cliff edge like other pueblitos, is not visible until one is within 40 meters or so of it. It could easily be missed, especially if one ascended to the bench along the Tapacito side of the confluence, where you need not come within a half a mile of the site to reach the confluence. Finally, if the pueblito was not a Navajo structure, but a Puebloan (Jemez?) family dwelling, as suggested by Towner (1996), political or kinship ties may have prevented some of the auxiliaries from informing the Spaniards.

The most parsimonious explanation for these differing data is, I suggest, that the pueblito was never seen by the auxiliaries or Spaniards. All of their activities took place nearer the confluence from which the structure is not visible. Therefore, Towner and Dean's (1992) interpretation that a hogan was built on the site in the summer of A.D. 1690 and the pueblito was constructed in A.D. 1694 remains the most viable interpretation of the site history.

Old Fort Ruin (LA 1869)

Old Fort is one of the largest and most complex pueblitos in the Dinétah. The site is on the north edge of a mesa, overlooking the east branch of San Rafael Canyon (Figure 4). Jacobson et al. (1992) indicate intervisibility with three smaller sites, LA 11089, 11187, and 11188, but Hidden House (LA 106573) and Rabbit House (LA 99804) are clearly visible as well. The site was originally described by Kidder (1920), and excavated by both Morris (Carlson 1965) and Keur (1944); additional research was recently conducted by Towner and Johnson (1998). Powers and Johnson (1987:14) provide a concise description of the site and indicate that it consists of eight forked-pole hogans and about twelve ground-floor rooms forming two room-blocks, all enclosed by a masonry wall (Figure 8).

Research at the site, by Kidder (1920), Morris (Carlson 1965), and Keur (1944) recovered faunal remains of domestic sheep, horse, dog, and deer, manos, arrowshaft smoothers, a metal ax, and a metate. The faunal remains represented a minimum of two sheep, one dog, two horses, and one deer. The ceramic assemblage represented at least 63 different vessels,

Figure 8. Homer Hastings's 1939 photograph of forked-pole hogans inside the courtyard at Old Fort Ruin. (Courtesy of James M. Copeland, Farmington BLM)

35 of which were Gobernador Polychrome. Unfortunately, Morris did not save many of the Dinétah Gray sherds or vessels.

Old Fort has played an important role in the interpretation of pueblitos since Kidder's (1920) work early in this century. Most notably, the site provided tree-ring-dated ceramic specimens, including Gobernador, Ashiwi, Puname, Tewa, Puyapki, and Hawikuh polychromes, Jemez Black-on-white, Rio Grande Glaze VI, Galisteo Glaze IV, and an abundance of Dinétah Gray (see Carlson 1965:11, Table 2 for a complete tabulation of the excavated ceramics). Dinétah Gray is the most common pottery found on the site today. Artifact descriptions provided by Carlson (1965) and Keur (1944), although probably skewed by their particular research interests, are undoubtedly more accurate than any current observations (e.g. Towner and Johnson 1998) because of the almost continuous surface collecting that has occurred at the site for the past 80 years. The majority of the ceramics were derived from two middens, one below Covered Entry # 1 and the other below Covered Entry # 2. Neither midden can be directly dated, however.

Architecture. Morris's original site map (Carlson 1965:9) and repeated visits to the site are used to infer a construction sequence independent of the tree-ring dates. The descriptions below are based on observations

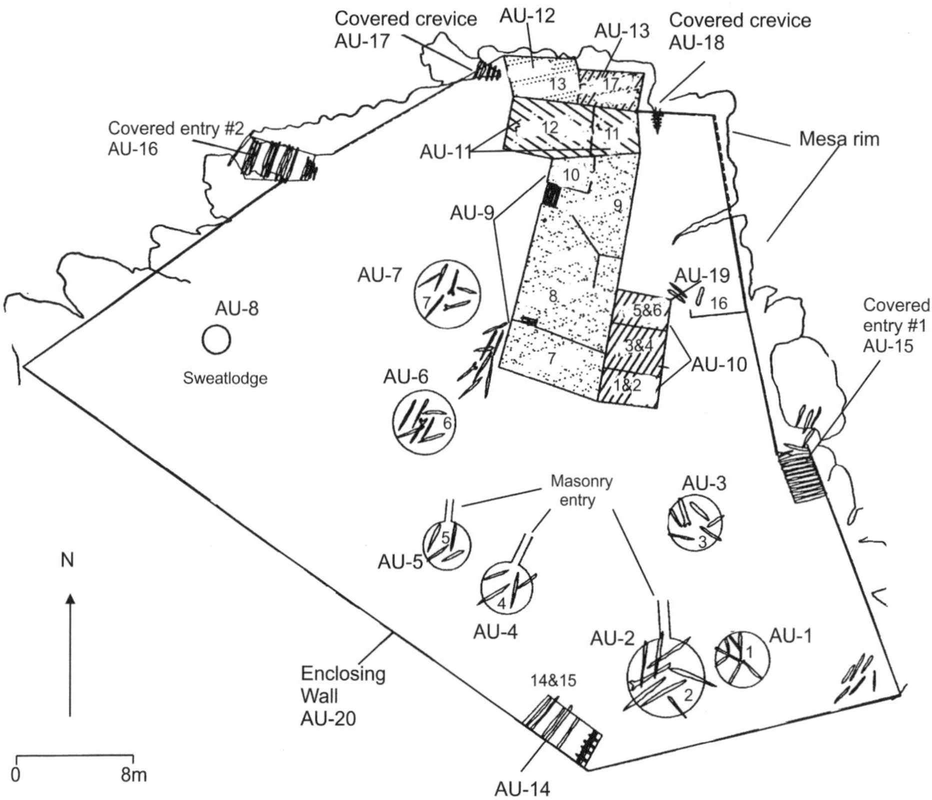

Figure 9. Map of rooms and architectural units (AUs) at Old Fort Ruin (LA 1869)

conducted during the collection of the tree-ring samples and are often different than Morris's 1915 site map. One of the hogans (# 8) is interpreted here as a sweatlodge because of a large rock pile nearby and the lack of structural remains. Other changes to the original map include the absence of the enclosing wall on portions of the north and northeast sides of the site, and the recognition of additional rooms (Rooms # 16 and 17) during the collection of the tree-ring samples.

There are twenty architectural units at Old Fort Ruin (Figure 9). The first eight units include Hogans 1 through 7 and the sweatlodge. Because all of these architectural units are spatially discrete, they cannot be related to other construction events based on architecture alone. Architectural Unit 9 includes Rooms 7, 8, 9, and 10 on the original site map. Although deterioration and wall fall have obscured many of the architectural details, the original site map indicates that these rooms were probably built as a unit. The outside wall surrounding these rooms is unbroken, except for the doorway on the northwest side, suggesting a single construction event. The interior walls dividing the space could have been added at any time, but *if* they are bonded to the outside wall, as indicated by the map, they must have been built at the same time as the exterior wall.

Architectural Unit 10 includes Rooms 1 through 6 on the original map. The east (exterior) wall of this unit appears to have been built as a single unit; again, *if* the interior walls are bonded to the exterior wall as indicated by the map, all six rooms were probably built at the same time. Based on the abutment of the west walls in these rooms onto the east wall of Architectural Unit 9, Architectural Unit 10 must have been constructed later than unit 9.

Architectural Unit 11 comprises Rooms 11 and 12 on the original map. These two rooms were apparently built as a unit, but their relationship to the other rooms is unclear. The original map suggests that the south wall of these rooms is bonded to the north wall of Architectural Unit 9. If this is the case, both architectural units must have been constructed at the same time. The change in angle of the south wall in Room 12, however, suggests that there is some temporal separation of the two units. The north wall of this unit appears, on Morris's map, to be the enclosing wall that surrounds the site. The large amount of rubble in this area of the site makes any evaluation of this relationship impossible. *If* Architectural Unit 11 is bonded to the enclosing wall and abutted to Architectural Unit 9, it suggests that both the enclosing wall and Rooms 11 and 12 were built at the same time or earlier than any of the other masonry architecture at the site except Architectural Unit 9.

Architectural Unit 12 comprises only Room 13. *If* the south wall of this room comprises part of the enclosing wall, it must have been built at the same time as that wall. Based on the abutment of the west wall of Room 17 onto the east wall of Room 13, Architectural Unit 12 must have been built before Room 17 (Architectural Unit 13).

Architectural Unit 13 comprises only Room 17. Depending on the relationship of the south wall and enclosing wall, this unit may have been built at the same time as the enclosing wall. It must have been built later than Room 13, however. Because Architectural Units 11, 12, and 13 cannot be absolutely contemporaneous, I believe the original site map is in error where it indicates these room walls bonded to the enclosing wall.

Architectural Unit 14 comprises Rooms 14 and 15 in the southern part of the site. These rooms are abutted onto the enclosing wall and must have been built later than that large unit. Architectural Unit 15 consists of Covered Entry # 1 on the east side of the site. Because this entry is an integral part of the enclosing wall, it is assumed to have been built at the time of wall construction. It is possible that the entry was added later, but there is no architectural evidence of remodeling to demonstrate such a sequence.

Architectural Unit 16 comprises Covered Entry # 2, a sealed crevice and trail leading up through the rock on the north side of the site. Because part of the enclosing wall rests on top of this construction, the entry must have been covered before the wall was constructed.

Architectural Unit 17 consists of a sealed crevice on the north side of the site. The map indicates the crevice must have been covered prior to wall construction because part of the enclosing wall rests on it. In reality, however, the enclosing wall does not exist in this part of the site. I assume that Morris either saw evidence of such a connecting wall, or inferred that the wall *should* have been present in this location.

Architectural Unit 18 is another small covered crevice east of Room 11. Again, the original map indicates that the enclosing wall rests on part of this construction; if that were the case, the crevice must have been covered before the wall was built. The enclosing wall does not exist in this area, however, and there is no evidence that it has collapsed or otherwise been destroyed.

Architectural Unit 19 consists of the scarce remains of Room 16. This room consists of only two small wall segments and the enclosing wall. The northern wall segment is abutted to Architectural Unit 10, indicating that it was constructed after Rooms 5 and 6. The southern wall segment is abutted to the enclosing wall, indicating that it was constructed after the wall was built.

Architectural Unit 20 consists of the enclosing wall. If the wall was built in a single construction episode, it must have been built after Covered Entries 1 and 2 were finished and before Rooms 14, 15, and 16 were added. The relationship between the enclosing wall and Architectural Units 11, 12, and 13 cannot be determined strictly from architectural evidence.

The architectural analysis indicates that Old Fort Ruin grew in several discrete stages. Unfortunately, I cannot infer the temporal place of any of the hogans or the sweatlodge in the site occupation. Architectural Unit 9, composed of Rooms 7, 8, 9, and 10, may have been the earliest masonry construction at the site. It is certainly earlier than Architectural Units 10 and 19, but its relationship to Architectural Unit 11 is unclear. The latter may have been built prior to Architectural Unit 9. In any case, the covered entries and covered crevice (Architectural Units 16, 17, and 18) must have been in place prior to construction of the enclosing wall (if the wall existed at Entry 3 and the crevice location) and Architectural Units 14 and 19 must have been added after the wall was complete.

Tree-Ring Samples and Dates. Previous tree-ring samples collected by several different researchers (Carlson 1965; Hall 1951; Robinson et al. 1974) confirmed a post-1700 occupation, and helped place both the site and the Gobernador Phase in time. Only one sample (RG-693) provided a cutting date (1749c), however, and no definitive construction episodes were defined, primarily because of the lack of provenience information for the tree-ring samples. The range of noncutting dates (1651++vv to 1748vv, with all but one date in the first half of the eighteenth century) did indicate an occupation prior to A.D. 1750. The most important of the samples (H-2) collected prior to 1992 was a v-cut taken from a tree living in the courtyard at the time of Morris's excavation in 1915. The earliest dated ring on this specimen is A.D. 1768 and it has been used to provide a terminal date for the site occupation (Carlson 1965:15), although it is not a pith date for the tree. Carlson assumed that the tree did not start growing until after the site was abandoned. Thus, prior to 1996, the most reasonable estimate was that the site was occupied from the 1720s until sometime prior to 1768.

Eighty tree-ring samples have been collected from Old Fort Ruin over the past 80 years (Table 3). Sixty-seven of these samples (85%) dated, one of Morris's samples is a duplicate, and fifteen (19%) yielded cutting dates. The samples collected by Earl Morris (n=11) (Carlson 1965) in the early part of this century lack provenience, but have provided a general temporal framework for the site since their publication in 1965 (Hannah 1965:109–13). Samples collected in 1992 and 1993 (n=69) are plotted on the site map (Figure 10).

Wood Use. Tree species used in the construction of the site reflect the local environmental distribution. Juniper (n=42 or 53%) and pinyon (n=33 or 42%) are the most abundant species used; only three Douglas-firs and a single ponderosa pine were part of the sampled specimens. Although Douglas-fir and ponderosa are available in the rincons around the San Rafael Canyon, pinyon and juniper would have been obtained much

Table 3. Tree-ring Samples Collected from Old Fort Ruin (LA 1869)

Sample Number	Species	Provenience	Inside Date	Outside Date	Terminal Ring	Tool Marks	Comment
DNT-163	PNN	Hogan 1 Fork	1608	1738+vv	inc	MA bark, limb, end, shape	flat
DNT-164	JUN	Hogan 1 Fork	NO DATE	—		MA bark, limb, end	
DNT-165	JUN	Hogan 1 Fork	1588	1732+vv	inc	MA limb, end, shape	flat
DNT-166	JUN	Hogan 1 Fork	1675	1747vv	inc	MA limb, end	
GP-3317	PNN	Hogan 1 (?)	1379p	1651++vv		Unknown	
DNT-169	JUN	Hogan 2 Fork	NO DATE	—		MA limb, end, shape	flat
DNT-170	JUN	Hogan 2 Leaner	1581	1746vv	inc	MA limb, end	
DNT-190	JUN	Hogan 3 Fork	NO DATE	—		MA limb, shape, ends	flat
DNT-191	JUN	Hogan 3 Leaner	NO DATE	—		MA limb, ends	
GP-3318	PNN	Hogan 3 (?)	1621+p	1741vv		Unknown	
GP-3319	PNN	Hogan 3 (?)	1609p	1748vv		Unknown	
DNT-201	JUN	Hogan 6 Fork	1635	1740+vv	inc	MA limb, end	
DNT-192	PNN	Hogan 7 Fork	1565	1719vv	inc	MA limb	
DNT-193	JUN	Hogan 7 Fork	NO DATE	—		MA limb	
GP-3323	PNN	Hogan 7 (?)	1475np	1722vv		Unknown	
GP-3315	PNN	Hogan (Unknown Number)	1581p	1732vv		Unknown	
GP-3316	PNN	Hogan (Unknown Number)	1526fp	1726vv		Unknown	
DNT-167	JUN	Bastion Corner	NO DATE	—		MA limb, end, shape	flat
DNT-168	JUN	Bastion Corner	1665	1745vv	inc	MA limb, end	
DNT-171	PNN	RM 14 Roof primary	1630	1752LB	inc	MA limb, ends	SB
DNT-172	JUN	RM 14 Roof primary	1649	1753v	comp	MA bark, limb, ends, shape	flat; SB
DNT-173	PNN	RM 14 Roof primary	1643	1744+vv	inc		SB
DNT-174	PNN	RM 14 Roof primary	1641	1746++G	inc	MA limb	
DNT-175	PNN	RM 14 Roof primary	1595	1752vv	inc	MA limb, ends	SB
DNT-188	JUN	RM 14 Roof primary	1564+/-	1751vv	inc	MA limb, ends	SB
DNT-189	PNN	RM 14 Roof primary	1627	1753v	inc	MA limb, ends	SB
DNT-176	JUN	Covered entry 1 Lintel	NO DATE	—		MA limb, end	
DNT-177	JUN	Covered entry 1 Primary	1502+/-	1746vv	inc	MA limb, ends	

DNT-178	JUN	Covered entry 1 Primary	1511+/-	1740+vv	inc	MA bark, limb, ends	
DNT-179	PNN	Covered entry 1 Primary	1647	1745vv	inc	MA bark, limb, ends	
DNT-180	PNN	Covered entry 1 Primary	1629p	1725vv	inc	MA limb, ends	
DNT-181	JUN	Covered entry 1 Primary	1701	1737vv	inc	MA bark, limb, shape, ends	split
DNT-182	JUN	Covered entry 1 Primary	1607	1706+vv	inc	MA limb, ends	
DNT-183	JUN	Covered entry 1 Loose log	1650	1746vv	inc	MA bark, limb, ends	
DNT-184	JUN	Covered entry 1 Door jamb	1659+/-	1750vv	inc	MA bark, limb, end	
DNT-185	PNN	Covered entry 1 Door jamb	1656	1745vv	inc	MA bark, limb, end	nocked
DNT-186	JUN	Covered entry 1 Door jamb	1639	1746+G	inc	MA limb, end	
DNT-187	JUN	Covered entry 1 Door jamb	1689	1746vv	inc	MA bark, limb, end	
DNT-194	JUN	Covered entry 3 Roof primary (?)	1649	1727vv	inc	MA bark, limb, end	
DNT-195	JUN	Covered entry 3 Roof primary	1536	1629vv	inc	MA limb	
DNT-196	JUN	RM 13 Loose primary	1627	1737+vv	inc	MA limb, ends	
GP-3322	PNN	RM 13 (?)	1407p	1742vv		Unknown	
DNT-197	JUN	RM 13 Roof primary	1581	1737+vv	inc	MA limb	
DNT-198	JUN	RM 11 Loose primary	1591	1739+G	inc	MA limb, ends	
DNT-199	PNN	RM 11 Corner Post	1660	1744+v	inc	MA bark, limb, end	
DNT-200	JUN	Covered crevice 1 Loose primary	1639	1745+vv	inc	MA limb, ends	
DNT-202	PNN	Covered entry 2 Roof primary	1555	1749G	inc	MA bark, limb, ends	
DNT-203	JUN	Covered entry 2 Roof primary	1666p	1750LB	comp	MA bark, limb, ends	
DNT-204	JUN	Covered entry 2 Roof primary	1614+/-	1721GB	inc	MA limb, ends	
DNT-205	PNN	Covered entry 2 Roof primary	1632	1730vv	inc	MA limb, ends	
DNT-206	JUN	Covered entry 2 Roof primary	1617+/-	1749v	inc	MA bark, limb, ends	
DNT-207	PNN	Covered entry 2 Roof primary	1428	1736+G	inc	MA limb, ends	
DNT-208	JUN	Covered entry 2 Roof primary	1655	1749B	inc	MA limb, ends	
DNT-209	PNN	Covered entry 2 Roof primary	1625	1737vv	inc	MA limb, ends	
DNT-210	PNN	Covered entry 2 Roof primary	1615	1746L	inc	MA limb, ends	
DNT-211	PNN	RM 9 Entry Vertical roof support	1682	1743vv	inc	MA bark, limb	
DNT-212	PNN	RM 9 Entry Vertical roof support	1674	1743L	inc	MA bark, limb	
DNT-213	JUN	RM 9 Entry Roof primary	1686	1746LGB	comp	MA limb, shape	flat
DNT-214	PNN	RM 9 Entry Roof primary	1654	1743L	comp	MA bark, limb	SB
DNT-215	PNN	RM 9 Entry Roof primary	1620	1743L	inc	MA limb	

Table 3 (Cont.). Tree-ring Samples Collected from Old Fort Ruin (LA 1869)

Sample Number	Species	Provenience	Inside Date	Outside Date	Terminal Ring	Tool Marks	Comment
DNT-216-1	PNN	RM 16 Loose primary	1633	1743vv	inc	MA limb, end	
DNT-216-2	JUN	RM 16 Loose primary	1665	1747vv	inc	Unknown	
DNT-217	JUN	RM 16 Loose primary	1545+/-	1750vv	inc	MA limb, end	
DNT-218	DF	RM 3/4 Loose primary	1684	1737vv	inc	MA end	
DNT-219	DF	RM 3/4 Loose primary	1659p	1741vv	inc	MA end	
DNT-220	PP	RM 3/4 Loose primary	NO DATE	—		MA bark, limb, end	broken end
DNT-221	JUN	RM 8 Door lintel	1645	1746G	inc	MA bark, limb, end	
DNT-222	PNN	RM 8 Door lintel	1591	1746vv	inc	MA limb, ends	
DNT-223	JUN	RM 8 Loose primary (?)	1641	1731vv	inc	MA bark, limb, end	
DNT-224	JUN	RM 8 Loose primary	1687	1747vv	inc	MA limb, ends	
DNT-225	JUN	RM 8 Vertical roof support	1634	1747vv	inc	MA limb, end	root flare
DNT-320	DF	RM 7 (?) Loose primary (?)	1688	1737vv	inc	MA end	
DNT-321	PNN	RM 7 (?) Loose fork (primary?)	1582p	1733++vv	inc	MA limb, ends	
GP-3321	PNN	RM 7/8 (?); same as RG-693	—	—	inc	Unknown	
RG-693	PNN	RM 7/8 (?); same as GP-3321	1565p	1749c	inc	Unknown	
DNT-322	JUN	Windbreak element	NO DATE	—		MA cut	
DNT-323	JUN	Windbreak Ax-cut limb	NO DATE	—		MA cut	
DNT-324	JUN	Windbreak element	NO DATE	—		MA cut	
GP-3320	PNN	Unknown masonry room	1412+p	1740vv		Unknown	
H-3	PNN	Unknown (Morris sample)	1341+p	1747vv		Unknown	
H-2	PNN	Living Tree cut by Morris in 1915		1915B		Unknown	

Key:
DNT= Tree-ring Lab Collection; H= Douglass Collection; GP= Gila Pueblo Collection; RG= Rio Grande Collection
JUN= Juniper; PNN= Pinyon; DF= Douglas-fir; PP= Ponderosa pine
MA limb= Metal ax limb trim; MA end= Metal ax-cut end; MA bark= Metal ax debarking; MA shape= Mteal ax shaping
SB= smoke blackening present; inc= incomplete; comp= complete

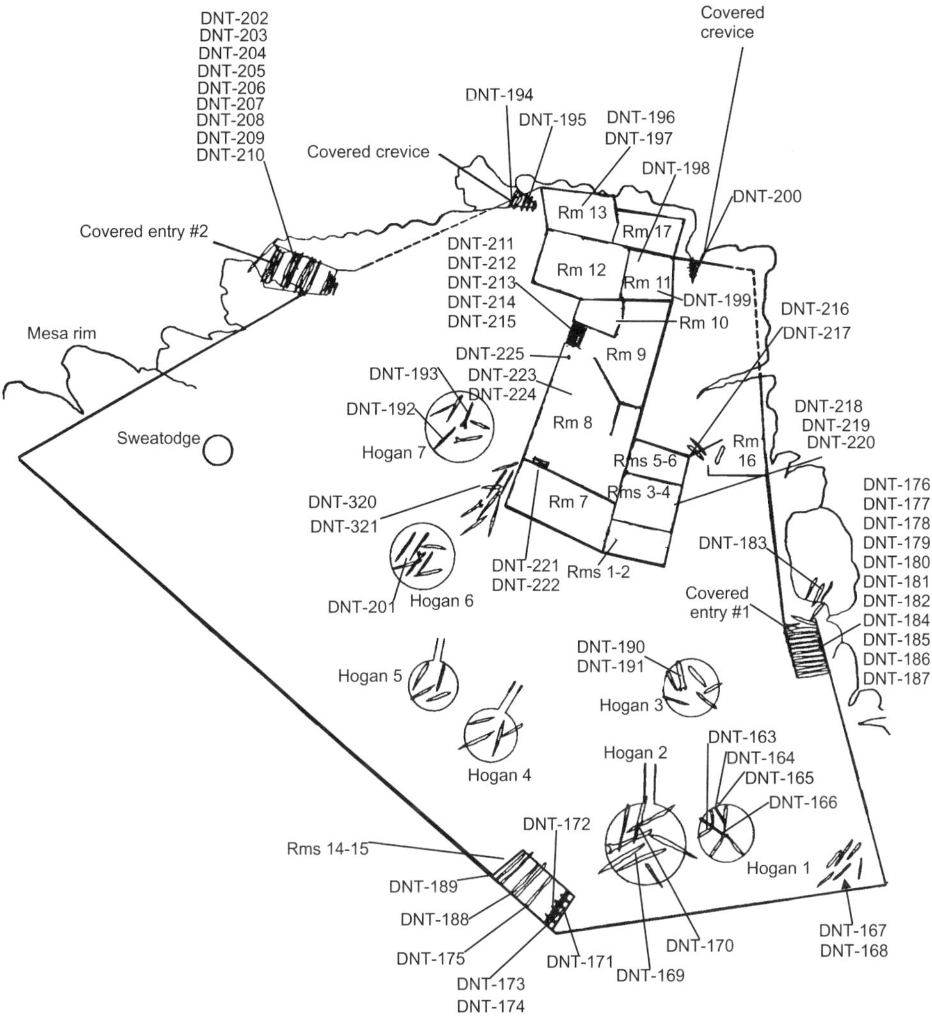

Figure 10. Tree-ring sample proveniences at Old Fort Ruin (LA 1869)

more easily nearer to the site. The nature of hogan construction allows the builders to select trees that may not be suitable for masonry construction (i.e., trees need not be straight and free of large branches). The somewhat informal masonry at Old Fort, however, does not appear to have required the long, straight, relatively branch-free beams that are necessary for more formalized site and room layouts. Rooms are abutted to each other and the roof spans are relatively short. These rooms apparently were constructed with less concern for roof size and structure shape than those found on Puebloan sites. At least one room (Room 7) had a cribbed roof. The less formalized nature of the Old Fort masonry probably allowed the builders to select wood based on other criteria, such as ease of procurement.

Tree-Harvesting Seasons. A cluster of cutting dates from a single provenience provides the strongest evidence of seasonality (Russell and Dean 1985; Towner 2002). A general distribution of terminal ring characteristics can also be used, but inferences based on such a distribution are not as strong as those from a cutting-date cluster distribution.

Twenty of the samples collected from Old Fort Ruin were considered usable for determining seasonality. Eleven of the samples are pinyon and nine are juniper. Ten of the pinyon samples have incomplete terminal rings, indicating their procurement during the growing season. A single pinyon has a complete terminal ring and was procured after the growing season of 1746. Although individual trees and years may vary somewhat, the pinyon growing season in northwest New Mexico generally begins in the late spring, usually June, and ends in late summer, usually late July or August.

The nine juniper samples with true outsides include three with complete terminal rings and six with incomplete terminal rings. The juniper growing season generally begins about the same time as the pinyon growing season (May or June), but extends later into the fall, often into September. The terminal ring characteristics of samples from the site indicate that most of the tree harvesting was probably conducted in the early summer. Only a few samples were procured later in the year. The small number of samples with true outsides, however, makes such an assessment of site seasonality somewhat tenuous.

Interpretations. Examination of the existing tree-ring dates simply as numbers can yield substantial information. The stem-and-leaf plot technique used by Ahlstrom (1985), indicates several episodes of tree harvesting (Figure 11). Although one cutting and two near-cutting dates are present prior to A.D. 1740 (1721GB, 1736+G, and 1739+G), most trees were cut in 1743 (n=4), 1746 (n=4), 1749 (n=4), 1750 (n=1), and 1753 (n=2). Other trees may have been cut in these same years, but the loss of outside rings on the samples makes the year of tree death impossible to determine.

Interpreting the dates in their archaeological contexts, however, provides a more detailed picture of the pueblito occupation. The earliest clus-

162	9
—	
165	1
—	
170	6
171	9
172	12567
173	01223677777789
174	00011233333445555566666666677777899 9
175	00012233

Figure 11. Stem-and-leaf plot of tree-ring dates from Old Fort Ruin (LA 1869)

ter of cutting dates comes from the covered entryway to Room 9 and dates to the late summer of 1743; this date cluster may not date the construction of the room, however. One of the beams in this entryway was cut in the fall of 1746; there is no architectural evidence to indicate that this is a repair element. If Architectural Unit 9 (Rooms 7, 8, 9, and 10) was built as a unit (as indicated by the architectural analysis), the 1743 dates must be anomalous. The doorway between Rooms 7 and 8 contains two dates (one cutting) at 1746. Room 8 also contains three other dated specimens; one loose roof primary beam dates 1731vv, another loose roof primary beam dates 1747vv, and an upright roof support beam dates 1747vv. Thus, it appears that Architectural Unit 9 was constructed in the summer or early fall of 1746 or later, and three of the entryway elements to Room 9 represent either dead-wood use or reuse of previously cut beams.

Architectural Unit 10 (Rooms 1–6) contains three sampled beams, only two of which dated. Both dates are noncutting dates from weathered roof primaries. The dates, 1737vv and 1741vv, indicate post-1741 construction, but the architectural unit must have been added after the construction of Architectural Unit 9 in 1746.

Architectural Unit 11 (Rooms 11 and 12) contains two sampled beams, both in Room 11. The loose roof primary yielded a near-cutting date of 1739+G, and the upright corner post yielded a near-cutting date of 1744+v. If this latter date accurately dates the construction of Architectural Unit 11, it further supports the inference that the 1743 dates from the Room 9 entryway are anomalous.

Architectural Unit 12 (Room 13) contains two sampled beams, both of which yielded noncutting dates at 1737. This architectural unit may have been the first masonry construction at the site, but this inference cannot be supported without additional work to determine the nature of the relationship between the enclosing wall and south wall in Room 13. It is somewhat unusual, but not impossible, that both beams yielded noncutting dates in the same year; both appear to have suffered substantial ring loss (as indicated by the "vv" symbol), however, and probably postdate 1737 by several years.

Architectural Unit 13 (Room 17) contained no suitable samples and can only be dated to a post–Architectural Unit 12 construction date (i.e., minimally post-1737).

Architectural Unit 14 (Rooms 14 and 15) yielded seven samples, all of which dated. Five of the samples were from primary roof beams and two were from secondary roof beams. Both of the secondary beams yield noncutting dates (1744+vv and 1746++G). Three of the roof primaries yielded cutting or near-cutting dates (1752LB, 1753v, and 1753v) and the other two roof primaries yielded noncutting dates (1751vv and 1752vv). The dates indicate that Architectural Unit 14 was built in 1753 with at least one primary beam that was cut during the growing season of 1752. The secondary beams date in the mid-1740s, although both are noncutting dates, and suggest reuse of these elements.

Architectural Unit 15 (Covered Entry 1) yielded 12 samples, 11 of which dated. Ten of the dates are noncutting dates that range from 1706+vv to 1750vv. Three of the noncutting dates are at 1746, as is the only near-cutting date (1746+G) from the structure. Using the guideline that the latest noncutting date from a structure (in the absence of a cutting-date cluster), may be used to infer a construction date (Haury 1935:105; Ahlstrom 1985:58), Architectural Unit 15 may date to A.D. 1750. The sample that dates at 1750 (DNT-184), however, is a "door jamb" that may have been added as a repair element. If that is the case, the structure probably dates to A.D. 1746.

Architectural Unit 16 (Covered Entry 2) yielded nine samples, all of which dated. All of the samples are from primary beams and yielded six cutting dates (1721GB, 1746L, 1749v, 1749G, 1749B, and 1750LB), one near-cutting date (1736+G), and two noncutting dates (1730vv and 1737vv). The cutting dates indicate that the structure was built in A.D. 1750, using freshly cut beams (n=1), beams that had been seasoned for a year (n=3), beams reused from other structures (n=2), and dead wood (n=1).

Architectural Unit 17 (Covered Crevice 1) yielded two samples, both of which yielded noncutting dates (1629vv and 1727vv). These dates indicate a post-1727 construction date, but the actual construction date is probably many years later.

Architectural Unit 18, the second covered crevice, yielded only one sample; it provided a noncutting date of 1745+vv. The sample has apparently suffered substantial ring loss, and probably postdates 1745 by several years.

Architectural Unit 19 (Room 16) yielded three samples, all of which dated. The three noncutting dates (1743vv, 1747vv, and 1750vv) are all from loose primary beams. The dates indicate construction of this room after A.D. 1750, which agrees quite well with the architectural analysis.

Architectural Unit 20, the wall enclosing the site, contained no samples to date its construction. If the wall was built as a single construction episode, that event must postdate the construction of the covered entries and the covered crevices (if it existed in those areas). Both Covered Crevice 1 and Covered Entry 2 suggest that wall construction occurred in or after A.D. 1750. Evidence suggesting a 1746 construction of at least part of the wall near Covered Entry 1 is derived from the cluster of 1746 noncutting dates in that structure. The presence of a door jamb dated to 1750, however, lends some support to the later construction date.

There are 17 samples and 12 dates from the hogans, but no cutting dates. Hogan 1 postdates A.D. 1747; Hogan 2 postdates 1746; Hogan 3 postdates 1748; Hogan 6 postdates 1740; and Hogan 7 postdates 1722 (probably by many years) (Table 3). Nine of the hogan samples are juniper and eight are pinyon; the number of pinyon samples is probably inflated, however, by the six pinyon samples collected by Gila Pueblo. The few tree-ring dates from the hogans suggest that they were probably occupied at the same time as the masonry structures.

Summary. The tree-ring and architectural data indicate that the initial construction at Old Fort Ruin occurred in 1746. The central room complex, consisting of Rooms 9, 11, and possibly 12, was completed at that time. Additional architectural units, such as Architectural Units 10 and 13, may have been built shortly after the core area of the site was completed. The site occupation included several additional architectural units that were completed after the initial construction was finished. The covered entries and covered crevices were probably completed in 1750, as was the enclosing wall. Architectural Unit 19 (Room 16) was added after

Figure 12. Truby's Tower (LA 2434)

1750. The final construction episode was the completion of Architectural Unit 14 (Rooms 14 and 15) in 1753.

One of the unique features of the site has been the presence of forked-pole hogans inside the compound wall surrounding the site. Inferring the contemporaneity of the hogans and masonry structures has, thus far, been done on the basis of spatial association. Although none provided cutting dates, tree-ring samples suggest that the hogans were contemporaneous with the masonry structures and provide much stronger evidence of a simultaneous occupation of both structure types.

The tree-ring dates indicate that the ceramics recovered by Morris (Carlson 1965: Table 2) were all in use in the 1740s and early 1750s. Unfortunately, the provenience information and short-lived nature of the site prevent me from making any definitive statements regarding the temporal ranges of various ceramics. It is interesting, however, that a single sherd of Hawikuh Polychrome was recovered from the area near Covered Entry # 2; previously thought not to have been produced after about 1700, one vessel (at least) was probably used at Old Fort in the late 1740s or early 1750s.

The occupation of Old Fort was relatively short-lived. The earliest construction occurred in 1746 and the latest date from the site is 1753.

Table 4. Boulder-top Pueblitos Used in the Analysis

Discussed	Not Discussed	
Three Corn Ruin	Boulder Fortress	Pointed Butte
McKean Pueblito	Canyon View	Kin Yazhi Pueblito
Truby's Tower	Jaramillo Canyon	Pork Chop Pass Site
Hill Road Ruin	Deer House	The Citadel
Kin Naa dáá	Crow Canyon Pueblito	Tso'bito Pueblito
Simon Canyon	Delgadito Pueblito	Tséatah Pueblito
Rabbit House	Dos Cerritos	42 Pueblito
Los Peñoles	Foothold Pueblito	Morris I
	Gomez Canyon Pueblito	LA 2305
	Split Rock Ruin	Santo Niño Canyon
	Dooyiit'íiní Pueblito	

Given the relatively continuous nature of tree harvesting activities after the initial construction episode at the site (cutting dates in 1746, 1749, 1750, and 1752), it is probable that any occupation later than 1755 would have left some tree-ring evidence. Carlson's estimate that the site was abandoned by 1768 is certainly correct, although it may actually overestimate the length of the site occupation. Thus, it appears that Old Fort Ruin, one of the largest and most complex pueblitos in Dinétah, was occupied for fewer than ten years.

BOULDER-TOP PUEBLITOS

A large group of sites included in the analysis consists of boulder-top pueblitos (Table 4). These sites include masonry structures, not simply wall segments, located on top of boulders. Some sites are positioned in valley bottoms, some are on benches, and some are atop large massifs (Figure 12). Some boulder-top pueblitos are situated on boulders adjacent to mesas (Boulder Fortress for example) and might be classified as mesa rim sites. Such sites are classified as boulder-top pueblitos because of the presence of masonry on detached boulders and not on the mesa rim. Almost all of these sites are associated with forked-pole hogans, either on the boulder, like Three Corn Ruin, or in the immediate vicinity (cf. Marshall 1991; Powers and Johnson 1987).

Three Corn Ruin (LA 1871)

Three Corn Ruin is one of the largest and most "defensible" pueblitos in the Dinétah. The site contains at least 13 rooms and possibly two hogans

Figure 13. Three Corn Ruin

located on a massive sandstone outcrop on a bench above the south side of San Rafael Canyon (Figure 4). The pueblito is not connected to the mesa top; this isolated pinnacle of rock is separated from the mesa above by several deep chasms. The sandstone massif is high above the surrounding benches and affords an excellent view of the surrounding terrain (Figure 13), including a large portion of San Rafael Canyon and areas beyond the traditional boundaries of Dinétah. Jacobson et al. (1992:135) list two other pueblitos visible from Three Corn, Old Fort (LA 1869) and LA 11089, but Rabbit House (LA 99804) and Kin Tsézéí (LA 11092) are probably visible as well. A well-used trail from the mesa top to the north side of the sandstone massif has probably been used since the site was occupied. Below and to the north of the site, broad benches give way to the San Rafael Canyon bottom approximately one kilometer distant. In addition to the structures on the sandstone outcrop, at least three sweatlodges, one slab-lined cist, a possible room or hogan foundation, and a large midden surround the pueblito (Towner and Johnson 1998).

Material Culture. Three Corn Ruin was investigated and partially excavated by Kidder (1920), Earl Morris (Carlson 1965), and Dorothy Keur (1944). Additional investigations, primarily tree-ring collections, have been conducted by the Navajo Land Claim personnel (Van Valkenburgh and Yazzie 1956), the University of New Mexico, and Towner and Johnson (1998). Faunal remains recovered represent four sheep, one elk, one mule deer, one dog, and at least two horses. The largest artifact assemblage from the site comprises European trade beads. More than 12,000 beads, comprising seventeen types were recovered from Burial 1. Most of these beads were manufactured in the eighteenth century, but a few may have been made as early as the mid-1600s (Carlson 1965:91). Pottery from the site, especially Puebloan tradewares, was described by Kidder (1920) and summarized by Carlson (1965:19, Table 3). The ceramics represent more than 88 different vessels, and were derived mostly from the northeast burial knoll (Carlson 1965:22). The collected ceramics are Gobernador Polychrome (n=36), Ashiwi Polychrome (n=11), other Navajo painted vessels (n=9), Puname Polychrome (n=6), Puyapki Polychrome (n=5), other Rio Grande wares (n=16), Anasazi wares (n=4), and one of the earliest Chinese porcelain plates recovered in New Mexico. Unfortunately, Morris did not collect Dinétah Gray sherds or vessels, so these tabulations cannot be used for comparative purposes. Three Corn Ruin is the only locale where several of these ceramic types (in particular the Ashiwi Polychrome vessels) have been tree-ring dated. It should be noted, however, that these vessels were recovered from the burial knoll, rerecorded by Towner and Johnson (1998) as LA 11096, and not the dated architectural units. The tree-ring dates, therefore, may not directly apply to the ceramics. The northeast burial knoll contained at least five burial pits; all of which were excavated by Morris in 1915. The ceramics recovered from the area, even if slightly misdated, are an important aspect of the occupation and, especially, abandonment of Three Corn Ruin. Carlson (1965:21) describes the northeast burial area as

> literally covered with fragments of broken pottery. It soon became apparent that this was not a chance assortment of sherds . . . but a purposeful deposit representing many broken, but restorable, vessels. . . . The deposit suggests that before Site 4 [Three Corn Ruin] was abandoned the inhabitants took such vessels as they had no intention of carrying and broke them on the surface of the principal cemetery. No pottery was found within the graves.

Architecture. The original site map (Carlson 1965) shows the site composed of thirteen rooms, two hogans, and a defensive wall surrounding

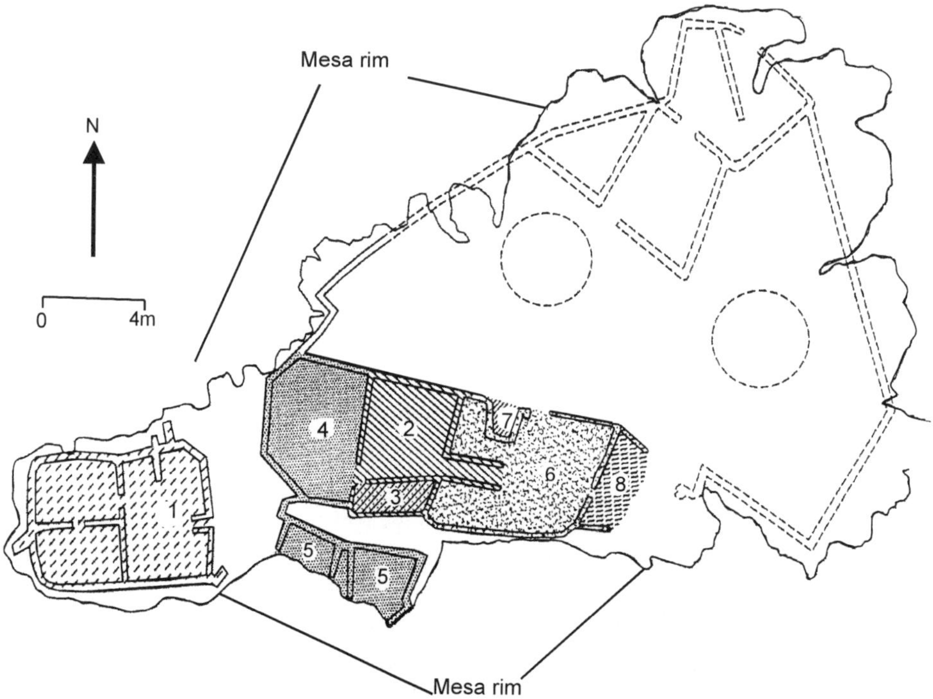

Figure 14. Architectural units at Three Corn Ruin (LA 1871). (After Carlson 1965)

the structures. Currently, the area of the hogans, Room 13 and the unnumbered rooms adjacent to Room 13, is buried in wall fall, rubble, and sagebrush. Neither the hogans nor the Room 13 complex can be adequately defined on the ground and, therefore, are not included in this analysis. There are several discrepancies between the previous maps and current observations, however, particularly in the area of Rooms 9 and 10. For reasons that will become obvious, I believe the original site map to have been an oversimplification of the relationships between the various rooms and construction events at the site.

The remainder of the site south and west of the hogans is relatively intact and provides much of the data concerning the site. Twelve rooms comprise eight architectural units in the southwest area (Figure 14). These architectural units can be further divided into an upper suite (Architectural Unit 1) and a lower suite (Architectural Units 2 through 8). The two suites are divided by sloping bedrock that has a series of hand and foot holds cut into it. Wall bonds and abutments are relatively clear in the upper suite, but become very complex in the lower room-suite.

Architectural Unit 1 is composed of Rooms 1 through 5. When the original site map was drawn by Morris, Room 3 comprised the second story of rooms 1 and 2 and Room 5 comprised the second story of Room 4. Unfortunately, these second-story rooms no longer exist. Rooms 2 and 4 are connected by a doorway, as are Rooms 2 and 3. Room 4 contains a small dividing wall on the north side that serves as a support for shelves. Carlson (1965:17) describes two storage bins below the shelves, but they no longer exist. It is apparent from the bonded walls that all of these rooms were built at the same time (although a second story certainly could have been added at a later time). The small dividing wall in Room 4 appears to be a later addition. Small wall stubs on the outside of the room walls on the west, north, and southeast sides appear to have been part of the original construction.

Architectural Unit 2 is composed of Room 7 and includes the covered entry into the room. It is the central room in the lower complex and was probably the first room built there. All of the existing walls that are parts of different rooms abut the walls of Room 7.

Architectural Unit 3 is the Room 7 annex. This room consists of a short wall segment extending from the east wall near the entrance on the south side of Room 7. The wall segment does not extend to the west wall of Room 7, and the semi-enclosed space, therefore, was labeled as a room annex. This architectural unit must postdate the construction of Room 7.

Architectural Unit 4 consists of Room 6. Only the southeast wall of this room remains intact, but there are small wall remnants indicating that it covered the area between Room 7 and the higher rock where the upper room-suite is located. The only adequate wall segment, on the southeast side, indicates that this unit postdates Room 7.

Architectural Unit 5 is composed of Rooms 8 and 12 on the south side of the lower suite. This complex is not connected to the other architectural units and cannot be relatively dated by architectural analysis. The north wall of this unit abuts the rock near the only extant Room 6 wall and may have been built at the same time, but such an inference should not be overstated.

Architectural Unit 6 consists of Room 9; it does not include the entryway into Room 7 that protrudes into Room 9. The construction history of Architectural Unit 6 is complex. The south wall of Room 9 abuts the south wall of Room 7, clearly indicating that Room 9 is the later of the two rooms. When Room 9 was added, a large masonry pillar was constructed on top of the existing Room 7 entryway to support the roof of

Room 9. Differences in masonry construction between the entryway and pillar clearly indicate that the pillar was added later. Probably at the same time as Room 9 was constructed, Room 10 was added to the north side.

Architectural Unit 7 comprises Room 10, a small storage area. Morris photographed a storage bin and a shelf in Room 10 (Carlson 1965:24, Plate 9a). The small shelf was constructed above the bins. The difficult aspect of the relationship between Rooms 9 and 10 is the roof. The roof of Room 10 is a sloping mortar and rock construction that joins the outer wall of Room 9. The roof of Room 9 actually is directly above the mortar roof of Room 10, making their temporal relationship impossible to determine. It is probable that the two rooms were constructed at the same time.

Architectural Unit 8 consists of Room 11. Although most of this room has deteriorated, it is clear that it was added after Room 9 was complete. The south wall of Room 11 abuts the east wall of Room 9 and must have been constructed later than Room 9.

The architectural analysis of the existing rooms at Three Corn Ruin indicates that the site was built in several different construction episodes. The two room-suites (upper and lower) comprise eight architectural units. Architectural Unit 1 (rooms 1–5) is isolated and cannot be temporally distinguished from the other architectural units. Architectural Unit 2 (Room 9) is the earliest room built in the lower suite. It was followed by Room 6 and Rooms 9 and 10, although which was added first cannot be determined from the architecture. Architectural Unit 8 (Room 11) was added after Room 9 was complete. The temporal relationship of Architectural Unit 5 (Rooms 8 and 12) to the rest of the architecture cannot be determined, although it may have been added at the same time as Architectural Unit 4 (Room 6).

Tree-Ring Samples and Dates. A total of 74 tree-ring samples has been collected from LA 1871; 66 of the samples (88%) yielded dates, more than half of which (n=38) are cutting dates (Table 5). Seventeen samples were collected prior to 1970 and are not well provenienced beyond the site level. The remaining 58 samples are provenienced to specific beams in and around the pueblito (Figure 15).

The range of dates (1583+vv to 1745cG) certainly indicates a historic period occupation; the range of cutting dates (1634L–1745cG) suggests a seventeenth- and eighteenth-century occupation. The temporal distribution of dates suggests an occupation beginning in the early 1700s (or before) and lasting until the 1740s (Figure 16). If the provenienced dates are

Table 5. Tree-ring Samples Collected from Three Corn Ruin (LA 1871)

Sample Number	Species	Provenience	Inside Date	Outside Date	Terminal Ring	Tool Marks	Comment
DNT-334	JUN	Roof primary RM 11	1610	1730+v	inc	MA bark, limb, shape, end	flat
DNT-325	JUN	Roof primary RM 9	1605	1730+GB	inc	MA limb, end	SB
DNT-326	JUN	Roof primary RM 9	1613p	1732GB	inc	MA limb, end	SB
DNT-327	PNN	Roof primary RM 9	1570	1733L	comp	MA limb, shape, end	flat; SB
DNT-328	JUN	Roof primary RM 9	1662	1732LGB	comp	MA limb, end	SB
DNT-329	JUN	Roof primary RM 9	1511+/-	1722++L		MA limb, shape	flat; SB
DNT-330	JUN	Roof primary RM 9	1565	1729+vv	inc	MA bark, limb	SB
DNT-331	JUN	Roof primary RM 9	1625	1730LB	comp	MA bark, limb, shape, ends	flat; SB
DNT-332	JUN	Roof primary RM 9	1610	1731L	inc	MA limb, shape, end	flat; SB
DNT-333	JUN	Roof primary RM 9	1558p	1729+LB	comp	MA limb, shape, end	flat; SB
DNT-335	JUN	Roof secondary RM 9	1648p	1734+v	comp	MA limb	SB
DNT-336	PNN	Roof secondary RM 9	1623	1732B	comp	MA limb	SB
DNT-337	JUN	Roof secondary RM 9	1669	1731B	comp		SB
DNT-344	JUN	Room 9/10 Above Passage	1670	1730GB	comp	MA limb, end	
DNT-345	JUN	Room 9 Above Passage	NO DATE	—		MA ends	split
DNT-338	PNN	Door lintel Passage North	1567	1728v	inc	MA bark, shape, ends	split; SB
DNT-339	PNN	Door lintel Passage North	1565	1728GB	inc	MA shape, end	split; SB
DNT-340	JUN	Door lintel Passage South	1629	1728LB	inc	MA limb	SB
DNT-341	JUN	Door lintel Passage South	1566	1729+B		MA limb, end	SB
DNT-342	JUN	Shelf Pole RM 10	NO DATE	—			split
DNT-343	JUN	Shelf Pole RM 10	1599	1723+v	comp	MA limb	
DNT-346	PP	Roof primary RM 7	NO DATE	—		MA limb, shape, end	flat; burned end
DNT-347	PP	Roof primary RM 7	1618	1726vv	inc	MA limb, shape, end	flat
DNT-348	PNN	Loose log RM 7	1596	1718vv	inc	MA limb, end, broken end	
DNT-349	JUN	Roof secondary RM 7	1680	1731G	inc	MA limb	

Table 5 (Cont.). Tree-ring Samples Collected from Three Corn Ruin (LA 1871)

Sample Number	Species	Provenience	Inside Date	Outside Date	Terminal Ring	Tool Marks	Comment
DNT-350	PNN	Roof secondary RM 7	1663	1732vv	inc	MA bark, limb	
DNT-352	JUN	RM 7 Alcove Roof primary	1633+/-	1731+vv	inc	MA bark, limb, end	SB
DNT-353	PP	RM 7 Alcove Roof primary	1639	1728v	inc	MA limb, shape, end	flat; SB
DNT-354	JUN	RM 7 Alcove Roof primary	1620	1734+G	comp	MA limb, end	SB
DNT-351	JUN	Loose Log RM 6	1620+/-	1728+B	inc	MA limb, end	SB
DNT-355	PNN	Roof primary RM 4/5	1606	1714v	inc	MA limb, ends	SB
DNT-356	PNN	Roof primary RM 4/5	1593	1714L	inc	MA limb, ends	SB
DNT-357	PNN	Roof primary RM 4/5	1641	1701G	inc	MA limb	no SB
DNT-358	JUN	Shelf Pole RM 4/5	1687	1736+LGB	comp	MA limb	SB
DNT-359	PNN	Shelf Pole RM 4/5	1692	1737v	inc	MA bark, limb	SB
DNT-360	PNN	Hatch seal RM 4/5	1637p	1722+vv	inc	MA limb	SB
DNT-361	PNN	Roof primary RM 2/3	1661	1714L	inc	MA limb	SB
DNT-362	JUN	Roof primary RM 2/3	NO DATE	—		MA limb	SB
DNT-363	JUN	Roof primary RM 2/3	1530	1634L	inc	MA limb	SB
DNT-364	PNN	Roof primary RM 2/3	1626	1709+LB	comp	MA bark, limb, ends	SB
DNT-365	—	Roof primary RM 2/3	LOST	—		MA bark, limb, ends	SB
DNT-366	PNN	Roof primary RM 2/3	1621p	1705L		MA limb, ends	SB
DNT-367	—	Roof primary RM 2/3	LOST	—		MA limb, ends	SB
DNT-368	PNN	Roof primary RM 2/3	1613	1709LGB	inc	MA limb	SB
DNT-369	PNN	Roof primary RM 2/3	1596	1705LB	comp	MA limb	SB
DNT-370	JUN	Roof primary RM 2/3	1583	1701+LB	comp	MA limb	SB
DNT-371	PNN	Roof primary RM 2/3	1633	1712L	inc	MA limb	SB
DNT-372	PNN	Roof primary RM 2/3	1618p	1711L	comp	MA limb, end	SB
DNT-373	JUN	Roof primary RM 2/3	1574	1656vv	inc	MA limb, ends	SB
DNT-374	JUN	Door lintel (east) RM 2/3	NO DATE	—		MA limb	no SB
DNT-375	PNN	Door lintel (south) RM 2/3	1648p	1708+v	inc	MA limb	no SB
DNT-376	PNN	Upright Roof Support RM 2/3	1530	1736L	comp	MA limb	no SB

DNT-377	PNN	Roof primary RM 1/3	1629+/-p	1714LB	inc	MA limb	no SB
DNT-378	PNN	Roof primary RM 1/3	1512p	1708+LGB	inc	MA limb	no SB
DNT-379	PNN	Roof primary RM 1/3	1616p	1713L	inc	MA bark, limb	no SB
DNT-380	PNN	Roof primary RM 1/3	1599	1706vv	inc	MA bark	no SB
DNT-381	PNN	Roof primary RM 1/3	1594p	1710G	inc	MA bark, limb	no SB
NLC-1004	PNN	Roof primary (?)	1622p	1720G		Unknown	
NLC-1005	PNN	Roof primary (?)	1624p	1728G		Unknown	
NLC-1006	PNN	Roof primary (?)	1647+p	1731+G		Unknown	
NLC-1007	PNN	Roof primary (?)	1664p	1724cG		Unknown	
NLC-1009	PNN	Hogan 4 (?) leaner	1454+p	1668+		Unknown	
NLC-1010	PNN	Hogan 4 leaner	1588+p	1707c		Unknown	
NLC-1003	PNN	Hogan 4 leaner	1621p	1715G		Unknown	
NLC-1008	PNN	Loose log South side	1511p	1712G	inc	Unknown	
NLC-1011	PNN	Loose log North side (?)	1570+p	1745cG		Unknown	
NLC-1012	PNN	Loose log North side (?)	1617p	1721G		Unknown	
GOV-1	PNN	Unknown	1593np	1728vv	inc	Unknown	
H-4	PP	Unknown	1363p	1583+vv	inc	Unknown	same as LAH-35-3, GP-3325
H-5	JUN	Unknown	NO DATE	—		Unknown	
RG-694	PNN	Unknown	1644p	1733+v	inc	Unknown	same as RG-695, GP-3326
RG-695	PNN	Unknown	—	—	inc	Unknown	same as RG-694, GP-3326
LAH-35-3	PP	Unknown	—	—	inc	Unknown	same as H-4, GP-3325
GP-3324	PNN	Unknown	1398p	1725+vv	inc	Unknown	
GP-3325	PP	Unknown	—	—		Unknown	same as LAH-35-3, H-4
GP-3326	PNN	Unknown	—	—		Unknown	same as RG-694, RG-695
GP-3327	PP	Unknown	1645p	1733+vv	inc	Unknown	
GP-3328	PNN	Unknown	1580p	1727v	inc	Unknown	

Key:

DNT= Tree-ring Lab Collection
GP= Gila Pueblo Collection
RG= Rio Grande Collection
NLC= Navajo Land Claim Collection
H= Douglas Collection

PNN= Pinyon
PP= Ponderosa pine
JUN= Juniper

MA limb= Metal ax limb trim
MA end= Metal ax-cut end
MA bark= Metal ax debarking
MA shape= Mteal ax shaping

flat= flattened
inc= incomplete
comp= complete
SB= Smoke blackening present

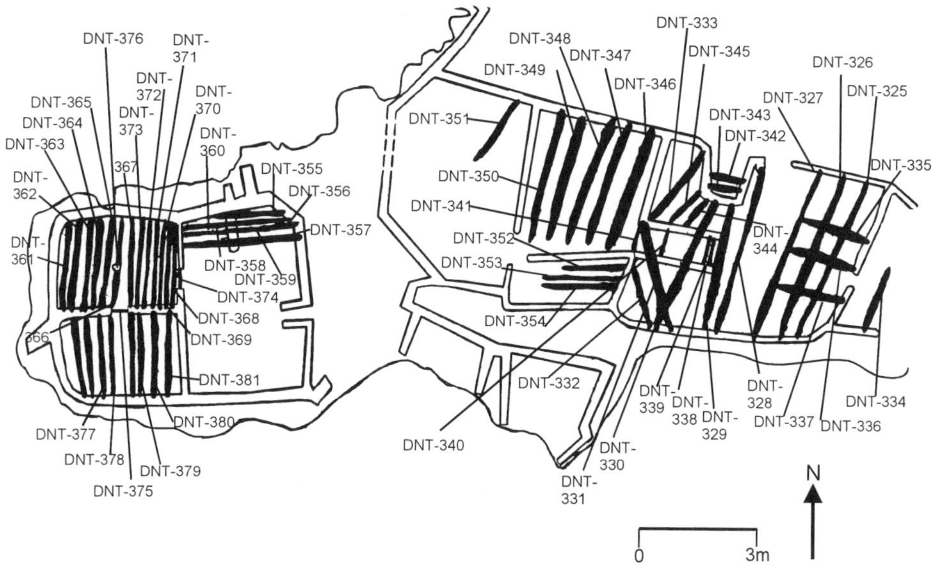

Figure 15. Tree-ring sample proveniences at Three Corn Ruin (LA 1871)

examined within their archaeological contexts, however, the occupation appears neither as long nor as continuous.

The earliest cluster of cutting dates is derived from Architectural Unit 1 in the upper suite of rooms. Twenty-seven samples were collected from Rooms 1/3, 2/3, and 4/5. The dates include cutting or near-cutting dates in 1634, 1701 (n=2), 1705 (n=2), 1708 (n=2), 1709 (n=2), 1710, 1711, 1712, 1713, 1714 (n=4), 1736 (n=2) and 1737. The earliest of these dates (1634) is clearly anomalous. The latest dates (1736 and 1737) clearly represent something other than initial construction at the site. The remaining dates, spread between 1701 and 1714, could be interpreted as construction and continuous repair of these rooms. It is possible that some of these beams were reused from a structure built in the first decade of the eighteenth century, but they do not date the construction of the upper roomsuite. Architecturally it appears that this suite of rooms was built in a single construction episode. If this is indeed the case, the suite of rooms could not have been built prior to 1714. Two of the roof primaries (of three sampled) in Room 4/5, the largest of the rooms, date to the summer of 1714; the other roof primary (DNT-357) dates to 1701 and is anomalous. Room 1/3 contains five sampled roof primaries; two noncutting dates in 1706 are accompanied by cutting dates in 1710, 1713, and 1714.

158 3

—

163 4

—

165 6

166 8

—

170 1155667899

171 01223444458

172 0122345678888888999

173 000011111222233344667

Figure 16. Stem-and-leaf plot of tree-ring dates from Three Corn Ruin (LA 1871)

This room was probably built in 1714 using freshly cut and previously used beams. Room 2/3 contains 10 dated roof primaries and a dated door lintel. The door lintel yielded a near-cutting date of 1708. The primaries include a noncutting date at 1656, and near-cutting dates in 1701 and 1709; other primaries include cutting dates at 1634, 1705 (n=2), 1709, 1711, 1712, and 1714. When all the architectural evidence and tree-ring data from the room-suite are combined, it is apparent that Room 2/3 was also built in 1714. Thus, it appears that the entire upper suite of rooms at the site was built in a single construction episode in the summer of 1714.

The lower set of rooms was not built until the late 1720s or early 1730s. Strictly on the basis of the dates, it appears that the passageway between Rooms 9 and 7 was the earliest construction episode in this part of the site. The passageway contains four lintels; three date to the summer of 1728 and the fourth to 1729. The passage shows no evidence of having been used as a storage area, and it is extremely unlikely that it was built prior to construction of Room 7. It appears to be an integral part of Room 7 and is similar to masonry entries on hogans and other structures at other pueblitos, for example, Old Fort Ruin and Hidden House. The passageway was modified by the addition of the masonry pillar on top of it after the passageway was completed.

Architectural Unit 2, which includes Room 7 and the passageway, yielded five samples in addition to those in the passageway. Only four of these samples dated, and only one provided a cutting date. Noncutting dates occur at 1718, 1726, and 1732; the only cutting date from the main part of Room 7 is 1731G. If the passageway is part of Room 7, as indicated by the architecture, construction of the room must have been in

1729 when the passageway was finished. It is probable that the original roof of Room 7 was removed and replaced at some time between 1731 and 1734; evidence for the 1734 reroofing date is present in the Roof 7 alcove, however, not in Room 7 itself.

Architectural Unit 3, the Room 7 alcove, yielded three samples suitable for dating. These three samples yielded a noncutting date of 1731+vv (DNT-352), a near-cutting date of 1734+G (DNT-354), and a cutting date of 1728v (DNT-353). These dates provide evidence for both the original construction date of Room 7 (1728–1729) and the reroofing event (1734) that created the Room 7 alcove.

Architectural Unit 4 (Room 6) yielded only a single sample, a loose primary roof beam that dated 1728+B. It is possible that this beam, currently lying loose in Room 6, was part of the original roof of Room 7. If that is the case, it lends some support to the interpretation of Room 7 as having been originally roofed when the passage was built and reroofed a few years later.

Architectural Unit 5 (Rooms 8 and 12) lacks any suitable samples for tree-ring dating. The temporal position of these rooms, therefore, remains unknown.

Architectural Unit 6 comprises Room 9. Fourteen samples were collected from this room, ten from primary roof beams and four from secondary roof beams. All ten primary beams from the roof of Room 9 yielded dates. Noncutting dates from the roof primary beams include those at 1722 and 1729; near-cutting dates include 1729+LB (DNT-333) and 1730+GB (DNT-325); cutting dates are all in the early 1730s and include 1730 (n=2), 1731 (n=2), 1732, (n=3), and 1733. Three secondary beams from Room 9 also yielded dates; cutting dates occur at 1731 and 1732, and a single noncutting date occurs in 1734.

The tree-ring dates from Room 9 indicate construction after the end of the pinyon growing season in 1733. Most of the primary roof beams were procured in the preceding three years, but were probably allowed to season prior to use. The 1734+v date from a secondary roof beam probably indicates minor repair of the roof after the end of the juniper growing season in that year. These dates agree quite well with those from Room 7 and the Room 7 annex, and with the interpretation that the latter rooms were reroofed when Room 9 was built.

Architectural Unit 7 consists of Room 10. Only two samples were collected from Room 10, both from shelf poles above the storage bins. Only one of these samples yielded a date, 1723+v (DNT-343). The architectural evidence indicates that this room must have been built during or after Room 9 construction; obviously, the single tree-ring date from this provenience is anomalous a cannot be used to precisely date room construction.

Architectural Unit 8 comprises Room 11. The only tree-ring sample from this room is a loose log that dates 1730+v (DNT-334). The room abuts Architectural Unit 6 (Room 9), however, and therefore must postdate 1733. The single date is clearly anomalous.

The tree-ring collection identified an additional construction event that cannot be discerned in architectural evidence alone. In the spring of 1737, the shelf and dividing wall were added to Room 4/5 using a pinyon (DNT-359) cut during the growing season and a juniper that had not begun growth in 1737. In either the fall of 1736 or early spring of 1737, the upright roof support was added to Room 2/3. It is probable that these events occurred at the same time as part of the reoccupation of the upper room-suite. Both construction events, even if separated by a few weeks or months, must have taken place after the end of the 1736 juniper growing season or early in the 1737 pinyon growing season. These minor rehabilitation events mark the final construction episodes at the site.

Three dates were derived from a hogan (Hogan # 4 in the NLC collections), although the relationship between the hogan and pueblito was not described when the collections were made. The dates include two cutting dates (1707c [NLC 1010] and 1715G [NLC 1003]) and a single noncutting date (1668+ [NLC 1009]). All the dates are from hogan "leaners," not forked poles, and indicate the hogan was occupied sometime after 1715. Whether or not it was occupied at the same time as the upper room-suite cannot be determined.

The unprovenienced samples from the site (n=14) provided the general framework upon which the site has been interpreted for the last few decades (Stokes and Smiley 1963:11). Although these dates generally follow the same distribution as the well-provenienced samples, there are a few exceptions. The single date that occurs in 1712 (NLC 1008) is from a loose log found on the south side of the ruin and may have been associated with construction of the upper room-suite. Most of the other NLC, Gila Pueblo, and Laboratory of Anthropology samples appear to have originated in the lower room-suite, although no proveniences are documented. Three dates, 1720G (NLC 1004), 1721G (NLC 1012), and 1724cG (NLC 1007) have no analogues in the provenienced samples and cannot be associated with any construction episode or architectural unit. The latest date from the site (1745cG) is derived from a loose log collected from the north side of the pueblito by the Navajo Land Claim personnel (NLC 1011). It appears to be an extreme outlier in the date distribution and is probably anomalous. It is improbable that the site was occupied for eight years after 1737, the final remodeling event in rooms 2/3 and 4/5, yet that occupation produced only a single tree-cutting event. This sample probably relates more to the general Navajo occupation of the area than to the occupation of Three Corn Ruin.

Wood Use. Tree species used in the construction of the pueblito and hogans are slightly different than at Old Fort (LA 1869) less than one mile away. At LA 1871, 39 of the sampled beams (53%) are pinyon, 27 (36%) are juniper, and five (7%) are ponderosa pine. No Douglas-fir beams were sampled (or noted) in the pueblito. These figures are slightly inflated because the Navajo Land Claim, Gila Pueblo, Laboratory of Anthropology (LAH and RG), and Morris sampling strategies excluded juniper. Nonetheless, it appears that the inhabitants of Three Corn Ruin did not select Douglas-fir or ponderosa pine for room construction despite their relative availability and superior quality for construction purposes when building rectangular rooms (see Towner and Dean's 1992 discussion of Tapacito Ruin [LA 2298] for a different example of species selection).

It may be significant, however, that the initial construction episode (Architectural Unit # 1) used primarily freshly cut and stockpiled pinyon and dead juniper. Only five of the twenty-one identified beams in Architectural Unit 1 are juniper, and all appear to have been used as dead wood. In contrast, the juniper beams used in the lower room-suite appear to have been cut at the same time as the pinyon beams.

Tree-Harvesting Seasons. The terminal ring characteristics of the beams used at Three Corn Ruin provide some insight into site seasonality. Beam clusters in several rooms indicate seasonal construction episodes. For example, although Architectural Unit # 1 was built with some stockpiled and some dead-wood beams, the freshly cut pinyon beams were all procured during the pinyon growing season in 1714. Likewise, both pinyon lintels and a juniper lintel in the Room 7 passageway were cut during the early summer of 1728, but not used until 1729.

In general, the terminal ring characteristics of all of the beams that can be assumed to have been cut specifically for use at the site (as opposed to dead-wood elements) suggest primarily spring and early summer tree-harvesting events. Thirty-four samples are appropriate for terminal ring analysis, eighteen pinyon, fifteen juniper, and a single ponderosa pine. The ponderosa pine was cut during the early spring, as were most of the pinyon trees (14/18). The juniper specimens were cut both during the growing season (n=8) and after tree growth had stopped (n=7). Based on the specific proveniences of the juniper samples with complete terminal rings, six of the seven samples were probably not used during the year they were cut. Most were probably used in later construction episodes and so may have been cut in the spring before the growing season began. Thus, most tree-harvesting activity at the site appears to have taken place during the spring and early summer.

Interpretations. The tree-ring evidence suggests that the site may have been built in two discrete episodes, one in the mid-1710s and the other in the late 1720s and early 1730s. The upper room-suite (and possibly Hogan 4) was apparently built as a single construction episode in 1714. The large number of pre-1714 cutting dates from this provenience, however, suggests that many of these beams were reused from an earlier structure (or structures) that no longer exists. A portion of the lower room-suite (Room 7) was constructed in 1728–1729; in 1732–1733, the roof of Room 7 was removed, the Room 7 annex was added, and Rooms 9 and 10 were constructed. Rooms 6 and 11 were added later, but there is no tree-ring evidence to indicate a specific year. A minor repair event in the roof of Room 9 occurred in 1734 and may be related to the addition of Rooms 6 and 11. The final construction episodes at the site were the addition of the shelf in Room 4/5 and upright support post in Room 2/3.

It is unlikely that the site (upper room-suite) was occupied throughout the late 1710s and early 1720s without at least some minor repairs. After the initial construction in 1714, no tree-cutting occurred until 1728. It is unlikely that the site was occupied for fourteen years without some tree cutting. Although I doubt the entire area was abandoned, the upper rooms at Three Corn Ruin certainly did not receive heavy use.

The entire site was probably abandoned shortly after 1737. The only post-1737 dated beam at the site is a loose log sampled by the Navajo Land Claim personnel. Again, given the relatively continuous nature of tree-cutting events from 1728 through 1737, I believe it unlikely that the site was occupied much later than 1737. This terminal date also indicates that those ceramics that have been dated by their association with Three Corn Ruin (Mera 1939), such as Ashiwi Polychrome, should not be assigned a terminal date of 1750.

Ethnohistory and the Abandonment of Three Corn Ruin. The archaeological and dendrochronological data suggest that the final abandonment of Three Corn was relatively sudden. The tree-ring dates suggest that the second occupation was relatively continuous from the late 1720s until sometime after A.D. 1737. The abundant ceramics from the northeast burial knoll (LA 11096) were not placed in the graves as burial goods, but as intentionally broken vessels deposited on the ground surface. Carlson (1965:21) suggests that the vessels were discarded because the inhabitants did not want to transport them to a new locale. This assumption ignores the burial area context of the vessels and seems untenable if that new locale were relatively close (i.e., Old Fort Ruin or some other nearby pueblito).

Ethnohistoric data, although somewhat ambiguous, may provide some insight into the final abandonment of the site. Van Valkenburgh (1940) relates a story about a battle with the Utes told to him in 1935 by Red Woman. The events she related occurred "two old men's lives ago. . . . That is 150 years of American counting" (Van Valkenburgh 1940:24). Although she had never been there, she described the Tower of Haskhek'izh (Spotted Warrior), as the headquarters of the chief of all the eastern Navajo and the location of a battle where the Ute lay siege to a "Navajo watchtower."

According to Red Woman, the Utes set out from a camp on the Los Pinos River, rode up the Largo, passed the watchtower at the mouth of Rincon Largo (LA 2433?), and attacked the site. Prior to the battle, the Navajos learned of the impending attack, and Spotted Warrior "made the Navajo work day and night to repair the tower and lay in supplies" (Van Valkenburg 1940:24). The battle and siege lasted 12 days, after which time a group of Navajos from Canyon de Chelly attacked the Utes, killing most and chasing the survivors north across the San Juan River. The Ute bodies were left unburied, and in later years medicine men used the Ute bones in ceremonies. Spotted Warrior abandoned the site and moved to Canyon de Chelly with his four wives and their children.

Red Woman said the site was located "five days hard ride" (presumably from her summer camp on Chinle Wash) on "the Mesa of the Hanging Pot," and could be identified by three petroglyphs carved under a ledge in the tower site: a jagged line, a St. Andrew's cross, and a jagged line with expanding ends. Van Valkenburgh's informants (1940:24) identified the petroglyphs as symbols for Male Lightning, a star, and for the Ant People, respectively.

Van Valkenburg later identified Hanging Pot (LA 2445), a site on a small mesa overlooking the headwaters of the Chaco River, as the site of the battle. He (1940:23) describes the site as containing a "half-dozen crumpled hogans . . . [and] a rock and log wall which skirted the sheer edge of the crag." Ceramics on the site included Acoma tradewares. There is certainly no reason to doubt the validity of Red Woman's account of the events surrounding the Ute attack. The period of the mid-eighteenth century is acknowledged as a time of conflict between the Navajo and Ute. Indeed, most pueblitos were probably built in response to Ute raiding pressures (Carlson 1965; Hogan 1992; Towner 1992, 1996). There may be reason to question Val Valkenburgh's identification of the location of the battle, however (Towner and Johnson 1998).

Hanging Pot (LA 2445) currently consists of a few scattered pieces of juniper and four small wall segments (see above). The petroglyphs have not been found despite an extended search (H. Rogers, personal communication). The wooden hogan elements are not arranged in any particular pattern, and the wall segments are only two or three courses high. The ceramics on the site are exclusively Dinétah Gray wares, and no tradeware pottery exists. There is little refuse either within the site boundaries or over the edges of the mesa. Five tree-ring samples collected from the site (see above) produced no cutting dates; three noncutting dates place the site occupation sometime in the 1740s. There are three additional hogans below the mesa, but the entire area appears to have been sparsely populated. The site exhibits no evidence of having housed a substantial number of people or of ever having been repaired. Compared to other pueblitos in the Dinétah, LA 2445 is best interpreted as short-term occupation by very few people.

Three Corn Ruin, on the other hand, is more likely to have been the "headquarters of . . . the chief of all the eastern Navajo." It is a relatively large site that was occupied for an extended period of time and shows evidence of extensive repair and remodeling. It also contains three petroglyphs (corn stalks) at its base, although they are not those described to Van Valkenburg. The ceramic and lithic materials on the site indicate extensive contact with peoples to the southeast, south, and west. The presence of charred timbers and burned adobe on the slope north of the pueblito led Powers and Johnson (1987:16) to suggest that at least portions of the site may have been burned.

If the tree-ring evidence suggests a sudden abandonment, Navajo tradition and the ceramic evidence from the northeast (LA 11096) and northwest (LA 11143) burial knolls provide a possible explanation for the abandonment (Towner and Johnson 1998). These sites contain five burials and hundreds of artifacts, all excavated by Earl Morris (Carlson 1965). Carlson provides no information concerning the sex or physical anthropology of the buried individuals, such as whether or not they died as a result of traumatic injuries. He does indicate, however, that the ceramics, including large quantities of Puebloan tradewares, were broken on top of the graves, not interred with the individuals. The sudden abandonment of Three Corn Ruin and the deposition of large quantities of ceramics on top of the burials may have both resulted from the death of one or more individuals in the building. Such behavior may have been a typical Navajo response to death, much as it remains today.

I am somewhat tentative in my suggestion that Three Corn Ruin was the locale of the events described by Red Woman. Hanging Pot fits the location described in the story, but seems unlikely in terms of its size, length of occupation, and lack of remodeling. Three Corn Ruin, on the other hand, does not fit the described location very well, but seems a much more likely site in terms of its size, duration of occupation, number of repair episodes, and associated site types. I believe that even if Three Corn Ruin was not the site of the events described by Red Woman, those events can be used as a model to interpret the abandonment of Three Corn Ruin.

McKean Pueblito (LA 112641)

McKean Pueblito is a ten-room masonry structure located on the east side of Blanco Mesa atop a 336 m^2 sandstone knoll between Big Rincon and Heifer Canyon in San Juan County, New Mexico, at an elevation of 2105 m (Figure 4). The knoll itself is located on a broad sandstone bench above both canyons, and stands approximately 15 m above the bench. The elevated topographic setting of the site provides an expansive viewshed, particularly to the east and along the Largo drainage system. No other structures are known from the area.

Nine of the rooms are oriented in two rows. Rooms 1–5 are located on the west side of the knoll, approximately 1 to 2 meters below the top of the knoll. Rooms 1–5 have associated wood beams, but only Room 4 has an intact roof. Rooms 6–10 are identifiable only by rock outlines, and none contain any roofing elements; they are directly on top of the knoll in a row adjacent to Rooms 1–5 (Towner et al. 2001: Figure 3). Room 10 is

Table 6. Tree-ring Samples Collected from McKean Pueblito (LA 112641)

Sample Number	Species	Provenience	Inside Date	Outside Date	Tool Marks	Comment
DNT-1017	PNN	RM 1 Loose log	NO DATE	—	Broken, burned ends	charrred
DNT-1018	JUN	RM 1 Roof primary	NO DATE	—	Broken end	SB
DNT-1019	JUN	RM 1 Loose log	NO DATE	—	MA end, broken end	weathered
DNT-1020	POP	RM 1 Roof primary	NO DATE	—	Broken ends	weathered
DNT-1021	PNN	RM 2 Loose log	1478	1692vv inc	None	weathered
DNT-1022	JUN	RM 3 Roof primary	NO DATE	—	MA end	weathered
DNT-1023	PNN	RM 3 Roof primary	NO DATE	—	Broken end	weathered
DNT-1024	PNN	RM 3 Loose log	1630+/-p	1707+vv inc	MA end	weathered
DNT-1025	JUN	RM 3 Roof primary	NO DATE	—	None	weathered
DNT-1026	JUN	RM 3 Roof primary	NO DATE	—	None	weathered
DNT-1027	PNN	RM 4 Roof primary	1633	1706vv inc	MA bark, limb, end	SB
DNT-1028	JUN	RM 4 Roof primary	NO DATE	—	Broken limb, MA end	SB
DNT-1029	PNN	RM 4 Roof primary	1505	1705+vv inc	MA shape, limb, end	SB
DNT-1030	PNN	RM 4 Roof primary	1571	1713B comp	MA shape, limb, bark, end	SB
DNT-1031	JUN	RM 4 Roof primary	NO DATE	—	Broken limb, MA end	no SB
DNT-1032	PNN	RM 4 Roof primary	1514	1708++v	MA shape, limb, bark, end	SB
DNT-1033	PNN	RM 4 Roof primary	1630	1713G comp	MA bark, limb	SB
DNT-1034	JUN	RM 4 Roof primary	NO DATE	—	None	fork
DNT-1035	JUN	RM 4 Roof secondary	1579	1707+B comp	None	no SB
DNT-1036	JUN	RM 4 Roof secondary	NO DATE	—	None	no SB
DNT-1037	PNN	RM 4/5 Door lintel	1644	1706+vv inc	MA limb	
DNT-1038	PNN	RM 4/5 Door lintel	1611	1707+vv inc	None	
DNT-1039	JUN	RM 5 Roof primary	NO DATE	—	MA limb	weathered
DNT-1040	PNN	RM 5 Roof primary	1610	1704+vv inc	MA shape (flat)	fork
DNT-1041	JUN	Ax-cut Stump	NO DATE	—	MA end	weathered
DNT-1042	JUN	Ax-cut Stump	NO DATE	—	MA end	weathered
DNT-1043	JUN	Ax-cut Stump	NO DATE	—	MA end	weathered

Key: DNT= Tree-ring Lab Collection; PNN= Pinyon; JUN= Juniper; POP= Cottonwood; limb= limb trim; bark= debarking; MA= metal ax; SB= Smoke blackening present

on the south side of the knoll and is not adjacent to the other rooms. It is associated with two circular stone features and a possible wall, but its function is unknown. Contrary to the evidence suggested by the site map, there are no identifiable bond-abut relationships for any of the rooms, and, therefore, the relative construction sequence for the rooms cannot be inferred.

Tree-Ring Data. Of the 27 samples collected (Table 6), 24 are from the pueblito itself, and three are from metal ax-cut stumps below the knoll. Towner et al. (1998) have suggested that culturally modified trees and stumps around early Navajo sites can provide important chronological information pertaining to the site occupations, but at McKean Pueblito this proved not to be true; none of the sampled juniper stumps dated because their erratic ring series do not match the local master chronology.

The species sampled at the site include 14 juniper (*Juniperus spp.*), 12 pinyon (*Pinus edulis*), and one cottonwood (*Populus spp.)* (Appendix). The pinyon and juniper trees were undoubtedly procured within a few hundred meters of the site—hence the presence of metal ax-cut stumps in the area. The cottonwood timber, however, must have been procured from a more mesic microenvironment, probably either a nearby spring or more likely from the valley bottom.

Eleven of the 27 samples collected (40.7%) yielded dates, but only two samples (DNT-1030, DNT-1033) provided cutting dates, and a single sample (DNT-1035) yielded a near-cutting date. The poor dating success rate of juniper (7.1%) is due to the opportunistic growth strategy of many junipers in this area. Pinyon, on the other hand, rarely produces false or locally absent rings, and therefore 10 of the 12 pinyons (83.3%) were crossdated. The single cottonwood sample (DNT-1020) lacks ring-width variability, contains only six rings, and was not crossdated.

The presence of many broken beam ends and limbs suggests that dead wood was procured for many architectural elements. Interestingly, none of the beams that exhibit broken ends or limbs produced dates. At McKean Pueblito, only Room 4 retains enough architectural integrity to yield information from smoke-blackened beams. In Room 4, six of the eight primary beams exhibit smoke blackening. Sample DNT-1034 lacks smoke blackening, probably because it is the outermost beam and has been exposed to weathering. As is discussed below, the absence of smoke blackening on beam DNT-1031, located in the middle of the room, has important implications for interpreting the tree-ring dates from the room.

Chronological Interpretations. The eleven dates from the site range from 1692vv inc to 1713B comp, but the only cutting dates cluster in 1713 and the only near-cutting date is 1707+B comp. The two 1713 samples (DNT-1030, DNT-1033) were cut after the 1713 growing season ended but before the initiation of growth in the spring of 1714. The sample dated 1708+B comp (DNT-1035) was cut between the end of the 1708 growing season and start of the 1710 growing season. The noncutting dates (1692vv–1707vv) lack evidence of true outsides and presumably have suffered some exterior ring loss due to weathering and erosion. As discussed below, however, the small cluster of noncutting dates in the first decade of the 1700s (DNT-1024, 1027, 1029, 1037, 1038, 1039, and 1040) suggests that exterior ring loss was minimal.

The date distribution can be interpreted in at least three different ways depending on the use of sample contexts. One possible interpretation is that the entire site was constructed in the fall of 1713 or spring of 1714 using a mixture of freshly cut timbers and possibly some dead wood. This interpretation is based on the fact that the latest dates, and only cutting dates (1713B comp), from the site are from in situ roof primary beams in Room 4. Dead-wood use may be indicated by sample DNT-1032 which dated 1708++v.

Several factors, however, suggest that this site interpretation may not be valid. If 1713 is the date of initial construction, all but one of the remaining samples must have consistently eroded between five and nine rings. I believe it is highly unlikely that erosional processes removed approximately the same number of rings from all of the remaining samples, particularly DNT-1024 (1707+vv) and DNT-1040 (1704+vv), which are located in different rooms. Finally, it must be remembered that the "vv" designation indicates that there is no evidence of a true outside on the sample; it does not indicate that there is unequivocal evidence that exterior ring loss has occurred (Robinson et al. 1974). Thus, the small cluster of dates in the first decade of the 1700s may actually indicate tree harvesting at that time, probably in 1707 or 1708.

Another possible interpretation is that the site was constructed in 1713/1714 with freshly cut timbers and reused wood from other structures or from dead trees. Sample DNT-1035 cannot have been cut in 1713/1714 and in this interpretation must be a reused beam. Sample DNT-1034 did not reveal an outside date, but it did have a forked end. The forked end suggests that it may have been used previously as part of a forked-pole hogan. Numerous other beams had at least one end broken, which could only be accomplished once the wood had been dead. Finally, the absence of any suitable structures in the area, combined with the abundant construction materials in the adjacent pinyon-juniper forest, indicate that beam reuse was minimal, if practiced at all.

A third possible interpretation of the tree-ring data is that McKean Pueblito was constructed in 1707 or 1708 using freshly cut timbers, and that Room 4 (at least) was remodeled in 1713 using freshly cut timbers and a piece of dead wood (DNT-1032). The inference of a 1707–1708 construction date is based on two 1707+vv dates, one from Room 4 and one from Room 3, the 1708++v date from Room 4, and the generally tight clustering of the 1705, 1706, 1707, and 1708 noncutting dates in Room 4. Although these samples do not show evidence of true outsides, I believe the clustering indicates tree harvesting in the same year. The provenience of the beams in Room 4 support this inference. All of the samples dated 1705–1708 are either roof secondaries or primaries located at either end of the room.

Interestingly, there is a large hole in the north wall of the structure that was initially thought to be a window. Closer analysis of field photographs, however, suggests that this hole—which opens onto Room 3—was created in order to facilitate the replacement of four primary beams (DNT-1030, 1031, 1032, 1033) under the secondary beams (DNT 1035, 1036) in the center of Room 4. Digital photographs show that this hole is irregular and retains neither upper nor lower window lintels. My interpretation is that because of the use of a different mortar type during the replacement event, the hole eroded out of the wall after the site was abandoned. These four beams include the two 1713 cutting dates, the piece of dead wood, and an undated sample; the undated sample may be an even later repair, as it is the only roof primary in the center of the room that does not exhibit smoke blackening. In general, I suggest the most parsimonious explanation of the dates and beam proveniences is that the structure was built in 1708 and Room 4 was remodeled in 1713.

Truby's Tower (LA 2434)

Truby's Tower is a five-room pueblito on a large boulder overlooking the central portion of Largo Canyon (Figure 4). Jacobson et al. (1992) indicate intervisibility between Truby's Tower, Tapacito Ruin, and Split Rock Ruin, but there is no such intersite visibility. An associated forked-pole hogan was reported previously, but there is no remaining evidence of such a structure.

Material Culture. Powers and Johnson (1987:32) indicate that a large collection for the Navajo Land Claim (Correll and Kurtz 1957) included mostly Dinétah Gray and Gobernador Polychrome ceramics, but Zia, Santa Ana, Zuni-Acoma, and Tewa wares were present as well. Farmer (1947) also noted both flaked and ground stone artifacts. Abundant material remains are still present at the site, despite years of authorized and unauthorized collecting.

Architecture. The pueblito consists of five or six rooms. The rooms on top of the boulder (Rooms 1–4) include two ground floor and two second story rooms, but the structure was apparently built in a single construction episode. Rooms 5 (lower) and 6 (upper) were adjacent to the southeast side of the boulder (Farmer 1947), but little remains of them. These two rooms may have been built as a separate construction episode. Powers and Johnson (1987:32) suggest that the upper rooms were reached via a roof hatch in Room 6, but no evidence of any such feature remains today. Several of the roof beams in Rooms 1 and 3 have been burned, but the uneven burning suggests a postabandonment fire.

Tree-Ring Samples and Dates. Thirty-five samples have been collected from the site, 14 by the Navajo Land Claim personnel, 19 as part of this project, and two by later stabilization crews (Appendix). These samples yielded 28 dates, including 8 cutting, 5 near-cutting, and 15 noncutting dates. The provenience information provided by the NLC samples, although more detailed than many other sites, may be in error. Specifically, the samples derived from "floor fill" in Room 2 are probably out of place. Room 2 is one of the second-story rooms, and it is more likely that these samples came from Room 1 or Room 3. In addition, the published proveniences (Stokes and Smiley 1969:9–10) for samples NLC-1323, 1324, 1326 and 1328 are slightly different from those written on the samples by the field collectors.

The species used at Truby's Tower include 30 pinyon, 4 Douglas-fir, and a single juniper tree. The frequent use of pinyon, although somewhat unusual, has been documented at several other pueblitos; the use of Douglas-fir is more common at pueblitos built in the 1740s and 1750s than those built earlier, but none of the Douglas-fir samples from this site dated. Fifteen of the samples were procured and/or modified using a metal ax. Ten specimens were burned, which I interpret as postabandonment burning.

Interpretations. If the architectural interpretation that the four rooms on the boulder were built during a single construction episode is correct, that event must have occurred in A.D. 1752 or 1753. The strong cluster of 1752 dates from Room 1 indicates that it was roofed during the pinyon growing season in 1752. None of the dates from the other rooms contradict this interpretation. It is possible that the 1753 date from the door lintel (DNT-523) actually dates the construction, but I consider this unlikely.

It is more probable that the door lintel was added after the building was complete. The noncutting 1709 and 1721 dates (DNT-532; NLC-1325) contribute little to the interpretation of Rooms 5 and 6. It is probable that these rooms were built at the same time as the main structure. Incomplete terminal rings are present on all specimens with true outsides and indicate tree procurement during the growing season, probably in the summer.

Hidden House (LA 106573)

Hidden House is a small five-room pueblito and associated forked-pole hogan located on the north side of San Rafael Canyon across from Old Fort Ruin (Figure 4). It is located on the first large sandstone bench above the canyon floor and is visible from Old Fort Ruin (LA 1869), and probably several other pueblitos as well.

Material Culture. Like those associated with many other pueblitos, the ceramic assemblage from Hidden House is dominated by Dinétah Gray wares and a few Gobernador Polychrome sherds. There is a panel of ceremonial imagery nearby that is clearly visible from the site. Hidden House has an unobstructed view of Old Fort, LA 11092, and LA 11153, but Three Corn Ruin and other pueblitos in San Rafael Canyon are not visible, although smoke from any of the pueblitos would have been easily visible from any other pueblito in the canyon.

Architecture. Hidden House consists of three masonry rooms on top of a large boulder, two masonry rooms at the base of the boulder, and a collapsed forked-pole hogan adjacent to one of the surface rooms. The entryway to the hogan may have been connected to one of the surface rooms. Although collapsed wall segments make interpretation difficult, there appear to be five architectural units at the site. Architectural Unit 1 consists of Rooms 1 and 2 on top of the boulder. Architectural Unit 2 comprises Room 3, which was abutted to Room 2. Architectural Unit 3 consists of Room 4, but there is no way to determine architecturally whether Room 4 was built at the same time as the boulder-top rooms. It is likely, however, that access to the boulder-top rooms was through the roof of Room 4. Room 5 (Architectural Unit 4) was added on to Room 4. The forked-pole hogan (Architectural Unit 5) was probably built at the same time as Room 4 and connected to it by a vestibule.

Table 7. Tree-ring Samples Collected from Hidden House (LA 106573)

Sample Number	Species	Provenience	Inside Date	Outside Date	Terminal Ring	Tool Marks	Comment
DNT-268	DF	Roof primary RM 1	1516	1617vv	inc	MA ends, limb, bark	flat
DNT-269	PNN	Roof primary RM 1	1573	1732vv	inc	MA ends, limb, bark	no SB
DNT-270	PNN	Roof primary RM 1	1665	1738L	inc	MA ends, limb	no SB
DNT-271	PNN	Roof primary RM 1	1608	1731vv	inc	MA ends, limb	no SB
DNT-272	PNN	Roof primary RM 1	1606p	1738L	inc	MA ends, limb	no SB
DNT-273	PNN	Roof primary RM 1	1689	1738v	inc	MA ends, limb	no SB
DNT-274	JUN	Closing Material RM 1	1546+/-	1732+vv	inc	MA shape	shingle?
DNT-275	PNN	Loose log RM 1	1666p	1737vv	inc	MA end, limb	no SB
DNT-276	JUN	Door lintel RM 1/2	1641	1737+L	inc	MA ends	SB
DNT-277	JUN	Door lintel RM 1/2	1594+/-	1737+LGB		MA ends	SB
DNT-278	PNN	Roof primary RM 2	1617	1733L	inc	MA ends, limb	
DNT-279	PNN	Roof primary RM 2	1589	1733vv	inc	MA ends, limb	
DNT-280	PP	Loose log RM 2	1655	1732+LG	inc	MA end, limb	same as DNT-290
DNT-290	PP	Roof primary RM 2	—	—	inc	None	same as DNT-280
DNT-281	PNN	Roof primary RM 3	1639p	1741vv	inc	MA end	SB
DNT-282	JUN	Roof primary RM 3	1636	1741L	inc	MA ends, limb	SB
DNT-283	JUN	Roof primary RM 3	1557	1724++GB		MA ends, limb	SB
DNT-284	PNN	Roof primary RM 3	1631	1737+vv	inc	MA ends, limb	SB
DNT-285	PNN	Roof primary RM 3	1540	1731vv	inc	MA ends, limb	SB; fork
DNT-286	PNN	Roof primary RM 3	1636p	1741vv	inc	MA ends, limb, bark	SB
DNT-287	PNN	Roof primary RM 3	LOST	—		MA ends, limb	SB
DNT-288	PNN	Roof primary RM 3	1619	1733vv	inc	MA end, limb	
DNT-289	JUN	Door lintel RM 3	1639	1740B	comp	MA ends, limb	SB; split

Key: DNT= Tree-ring Lab Collection; PP= Ponderosa pine; JUN= Juniper; DF= Douglas-fir; PNN= Pinyon; MA= Metal ax; bark= debarking; limb= limb trim; inc= incomplete; comp= complete; SB= Smoke blackening present

Tree-Ring Samples and Dates. Twenty-seven tree-ring samples were collected from the site, 21 from the pueblito and 6 from the hogan (Towner 1997). These samples produced 22 dates, 14 of which are cutting or near-cutting dates (Table 7). Juniper, pinyon, and Douglas-fir trees were used in construction. Juniper was used most frequently and comprises fifteen pueblito samples and four hogan samples; pinyon samples include five from the pueblito and two from the hogan; the single Douglas-fir sample is from a door lintel in Room 1. All but three of the beams show evidence of having been procured or modified using a metal ax.

Interpretations. The tree-ring dates indicate that the initial occupation of the pueblito began with the construction of Rooms 1 and 2 on top of the boulder, probably after the end of the A.D. 1746 growing season. Room 4 at the base of the boulder and the forked-pole hogan were added in 1751, probably as a single construction event. Although only a single date is available from it, Room 3 may have been added at the same time. Room 5 was added after Room 4 was complete. The roof of Room 4 was repaired by the addition of an upright support post during or after A.D. 1754, and the site was abandoned shortly thereafter.

Dead wood was used, but sparingly—only beams DNT-661 in Room 4 and DNT-643 Room 1 are clearly dead-wood elements. Reuse of beams, however, appears to have been common; five of the beams in Room 4 (DNT-656, 657, 658, 659, and 660) appear to have been reused from another structure, and two of the hogan upright entry poles were apparently reused. It is interesting to note, however, that the hogan forks were cut fresh in 1751. There is little evidence of the stockpiling of beams, although reuse of beams may mask evidence of stockpiling; only beams DNT-644 and DNT-646 in Room 1 appear to have been cut and stockpiled for future use.

Towner and Johnson (1998) indicate that the significance of the Hidden House tree-ring dates is that they correspond very well with events at Old Fort Ruin. The initial construction in 1746, site expansion in 1751, and abandonment after 1754 all correlate with similar events across the valley at Old Fort. Indeed, it is possible that some of the reused timbers at Hidden House were originally used at Old Fort, but such reuse cannot be demonstrated. It is clear, however, that Hidden House was closely tied to the Old Fort community and was unassociated with the previously abandoned Three Corn community.

Hill Road Ruin (LA 55833)

Hill Road Ruin is a four-room structure on top of a boulder overlooking Cereza Canyon (see Figure 4). Visibility is generally restricted northward down a small side drainage and into Cereza Canyon; Jacobson et al. (1992) indicate no other visible sites connected to Hill Road, but I believe there is at least one other unrecorded pueblito visible across the Cereza drainage. The structure is located part way down the hillslope overlooking the canyon (Figure 17).

Material Culture. Powers and Johnson (1987:104) identified Dinétah Gray ceramics, a few pieces of bone, and some flakes as the only artifacts on the site surface. Substantial midden areas surround the boulder and human remains are present in the general area, however.

Figure 17. Hill Road Ruin (LA 55833)

Architecture. This site shows one of the clearest construction sequences of any pueblito. The masonry structure was built in three construction episodes (Figure 18). Rooms 1 and 2 were constructed as a unit, Room 3 was abutted to Room 2, and Room 4 was later added to Room 3.

Tree-Ring Samples and Dates. Twenty-three tree-ring samples, including one duplicate, were collected from the site (Table 8). All of the samples produced dates, except a sample that was lost after collection. Most of the samples (n=13) are pinyon, some (n=6) are juniper, and both Douglas-fir (n=1) and ponderosa pine (n=1) are present as well. All of the beams exhibit tool marks made with a metal ax.

Interpretations. The first construction episode at the site, the building of Rooms 1 and 2 as a unit, occurred during the pinyon and juniper growing season in 1738, probably in the early to mid-summer. Predominately freshly cut beams were used in the construction. Room 3 was added in the spring or early summer of 1741; most beams had started growing, but one juniper (DNT-289) dated 1740B had not initiated growth in 1741. Again, most beams were freshly cut, although a single juniper (DNT-283) represents dead-wood use. Dendrochronological analysis cannot date the addition of Room 4 except to indicate that it was built after the summer of 1741.

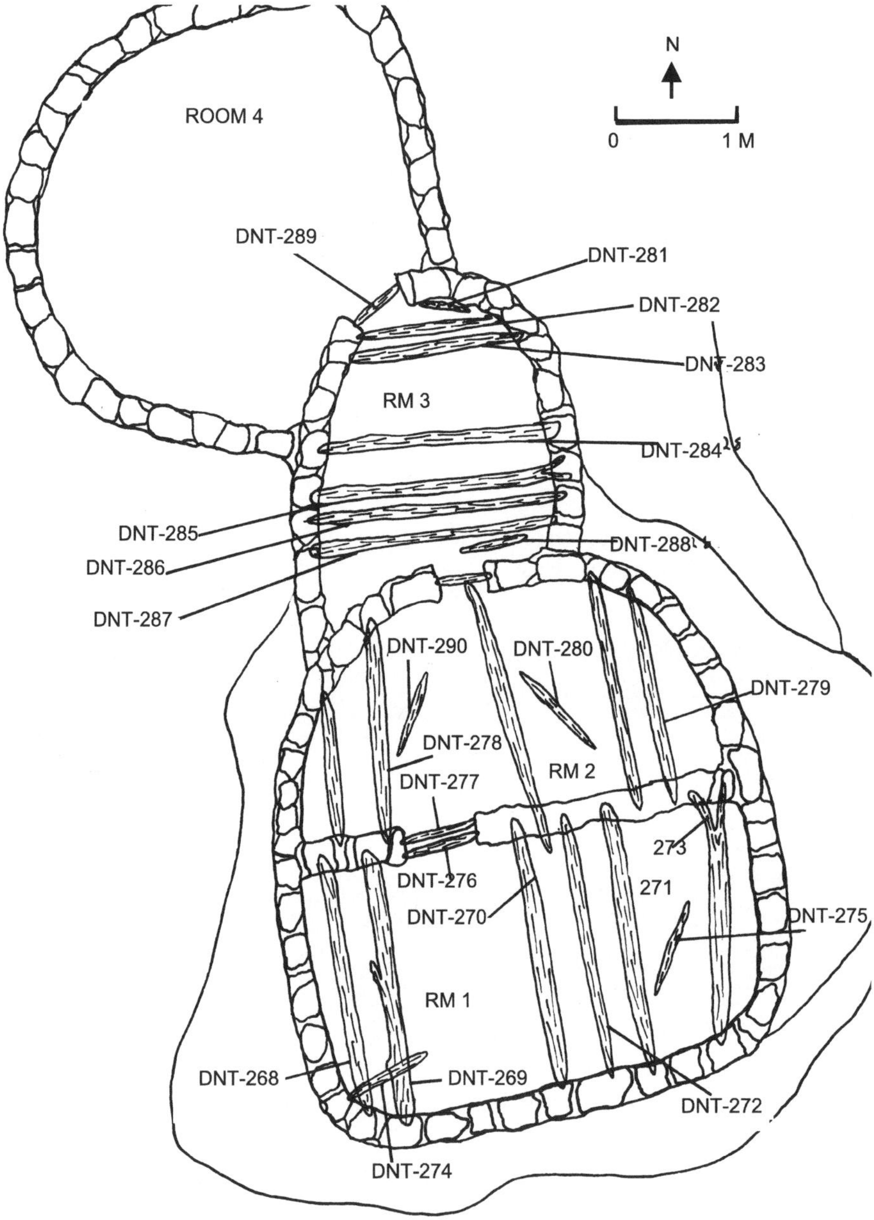

Figure 18. Tree-ring sample proveniences at Hill Road Ruin (LA 55833). (After Powers and Johnson 1987)

Table 8. Tree-ring Samples Collected from Hill Road Ruin (LA 55833)

Sample Number	Species	Provenience	Inside Date	Outside Date	Terminal Ring	Tool Marks	Comment
DNT-643	JUN	Roof primary RM 1	1521	1730++v	inc	MA bark, limb, end	
DNT-644	JUN	Roof primary RM 1	1598	1745G	inc	MA bark, limb, ends	
DNT-645	JUN	Roof primary RM 1	NO DATE	—		MA bark, limb, end	
DNT-646	JUN	Roof primary RM 1	1630	1744+v	inc	MA bark, limb, ends	
DNT-647	JUN	Closing Material RM 1	NO DATE	—		Broken ends	
DNT-648	DF	Door lintel RM 1	NO DATE	—		MA ends	
DNT-649	PNN	Door lintel RM 1	1684	1746B	comp	MA bark, limb, shape, ends	flat; spiral
DNT-650	JUN	Roof auxilary RM 2 (corner)	1635	1725vv	inc	MA ends	
DNT-651	PNN	Roof auxilary RM 2 (corner)	1652	1701vv	inc	Broken ends	
DNT-652	JUN	Loose log RM 3	—	—		MA bark, limb, shape, ends	same as DNT-653
DNT-653	JUN	Loose log RM 3	1643	1750+B	inc	MA bark, limb, shape, ends	same as DNT-652
DNT-654	JUN	Door jamb RM 4	1709	1754vv	inc	MA limb	
DNT-655	JUN	Auxilary cribbing RM 4	1610	1751+B	inc	MA limb, ends	fork
DNT-656	PNN	Auxilary cribbing RM 4	1640	1747G	inc	MA bark, limb, end	
DNT-657	JUN	Auxilary cribbing RM 4	1674	1743G	inc	MA bark, limb, end	
DNT-658	JUN	Auxilary cribbing RM 4	1674	1740+v	inc	MA bark	SB
DNT-659	JUN	Horizontal intramural RM 4	NO DATE	—		MA limb, end	
DNT-660	JUN	Corner Crib RM 4	1618	1747+G	inc	MA ends	
DNT-661	JUN	Corner Crib RM 4	1562	1714++vv	inc	MA end	
DNT-662	PNN	Corner Crib RM 4	NO DATE	—		MA limb, end	
DNT-663	JUN	Corner Crib RM 4	1676	1745vv	inc	MA end	
DNT-664	PNN	Door lintel RM 4	1627	1744+v	inc	MA end, broken end	
DNT-665	PNN	Hogan upright door entry	1656	1746G	inc	MA limb, end	
DNT-666	JUN	Hogan upright door entry	1658	1747+G	inc	MA limb, end	
DNT-667	JUN	Hogan horizontal door piece	1636	1735+vv	inc	MA limb, end	
DNT-668	JUN	Hogan forked pole	1614	1751G	inc	MA bark, limb; burned end	
DNT-669	JUN	Hogan forked pole	1652	1751+G	inc	MA limb, burned end	
DNT-670	PNN	Hogan leaner	1691	1746vv	inc	MA limb, broken ends	

Key: DNT= Tree-ring Lab Collection; PNN= Pinyon; JUN= Juniper; DF= Douglas-fir; PP= Ponderosa pine; MA= Metal ax; Limb= limb trim; bark= debarking; inc= incomplete; comp= complete; SB= smoke blackening present

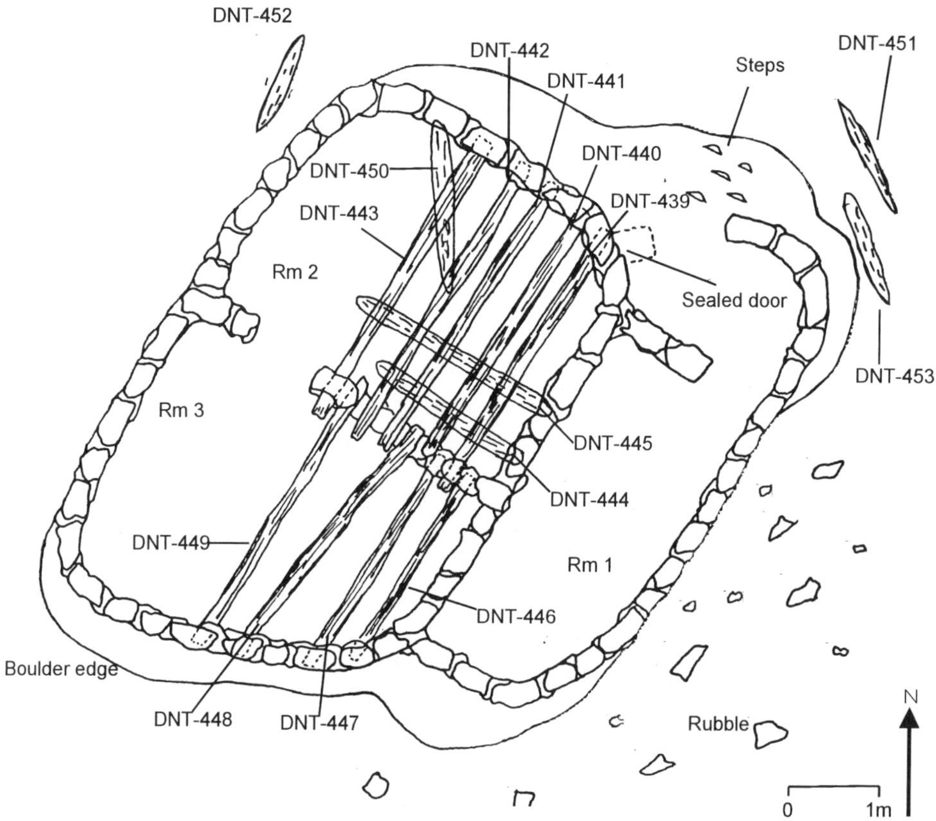

Figure 19. Tree-ring sample proveniences at Kin Naa dáá' (LA 1872). (After Powers and Johnson 1987)

Kin Naa dáá' (LA 1872)

Kin Naa dáá' is a three-room masonry structure on a small boulder near the valley bottom in a tributary of Gobernador Canyon (Figure 4). Jacobson et al. (1992) indicate that the site is visible from the Cabresto Mesa Tower Complex. The site was excavated by Morris in 1915 and originally had two additional second-story rooms (Carlson 1965:29–31).

Material Culture. Powers and Johnson (1987:19) and Carlson (1965:31) indicate the presence of Dinétah Gray, Gobernador Polychrome, Payupki Polychrome, and Ashiwi Polychrome ceramics, but do not indicate proportions. No data on other artifact classes are available.

Table 9. Tree-ring Samples Collected from Kin Naa dáá' (LA 1872)

Sample Number	Species	Provenience	Inside Date	Outside Date	Terminal Ring	Tool Marks	Comment
DNT-439	PNN	Room 2 Roof primary	1584	1726v	inc	MA limb, end	
DNT-440	PNN	Room 2 Roof primary	1628p	1727v	inc	MA bark, limb, end	
DNT-441	PNN	Room 2 Roof primary	1610	1727v	inc	MA limb, ends	
DNT-442	PNN	Room 2 Roof primary	1610	1727v	comp	MA limb, ends	
DNT-443	PNN	Room 2 Roof primary	1609p	1727v	inc	MA limb, end	fork
DNT-444	PNN	Room 2 Roof secondary	1616p	1727vv	inc	MA limb, end	
DNT-445	PNN	Room 2 Roof secondary	1605	1727v	inc	MA bark, limb, ends	
DNT-450	JUN	Room 2 Roof secondary (?)	1640	1725+vv	inc		
DNT-446	PNN	Room 3 Roof primary	1615	1727vv	inc	MA bark, limb, ends	
DNT-447	JUN	Room 3 Roof primary	1609	1725+B	comp	MA limb, ends	
DNT-448	JUN	Room 3 Roof primary	1613	1726vv	inc	MA limb, ends	
DNT-449	PNN	Room 3 Roof primary	1632p	1727v	inc	MA bark, limb, ends	
DNT-451	JUN	Loose log	1583	1721vv	inc	MA limb, ends	
DNT-452	PNN	Loose log	1567	1727v	inc		
DNT-453	JUN	Loose log	1604	1720vv	inc	Burned ends	
RG-696	PNN	Unknown	1630p	1727r	inc		

Key: RG= Rio Grande Collection; DNT= Tree-ring Lab Collection; PNN= Pinyon; JUN= Juniper; comp= complete; inc= incomplete; MA= Metal ax

Architecture. The four-room core of the structure, including the two upper-story rooms, appears to have been built as a single construction episode (Figure 19). A sealed doorway adjacent to the Room 1 entryway led Powers and Johnson (1987:19) and Carlson (1965:29) to suggest that Room 1 was added onto the existing structure and that access was gained through its second story. Although the second story has collapsed, this explanation of the site architecture still seems to be plausible.

Tree-Ring Samples and Dates. Sixteen samples have been collected from the site, including 12 from in situ beams (Table 9). All of the in situ beams, with the exception of a roof secondary in Room 2, show evidence of metal-ax modification or procurement. Pinyon appears to have been specifically selected for roof beams (n=9), although juniper (n=3) was also used.

Interpretations. The tight clustering of cutting dates from the site and all sampled rooms, combined with the architectural analysis, indicates construction as a single event in the late summer of A.D. 1727. Although most of the 1727 samples that yielded cutting dates retain incomplete terminal rings, the single complete terminal ring (DNT-442) indicates that this particular tree had ceased growing for the 1727 season. None of the samples postdate 1727, and thus we cannot date the addition of the Room 1 entryway with any degree of precision. The lack of repair events suggests that the site was occupied for only a short time.

The 1727 dates suggest Kin Naa dáá' may have been part of a larger community that included the second occupation at Three Corn Ruin (LA 1871) in San Rafael Canyon (Towner and Johnson 1998). It is also possible that Romine Canyon Pueblito (LA 55836), located across the Gobernador drainage, was occupied at the same time (Dykeman and Wharton 1996).

Simon Canyon (LA 5047)

Simon Canyon is a small, one-room structure on top of a large boulder. The boulder is on the east bank of Simon Canyon (Figure 4) on the first terrace above a spring (Powers and Johnson 1987:37). No other pueblitos are visible from Simon Canyon. It is the most northwesterly of all pueblitos and the only dated example north of the San Juan River. It is the only site surveyed by Marshall (1991) that was not associated with other sites or features, although there is a large hogan village above the canyon to the northwest.

Material Culture. Powers and Johnson (1987) noted a light scatter of Dinétah Gray and Gobernador Polychrome sherds and a few flakes on the bench, but no midden is associated with the site (Marshall 1991).

Architecture. The site consists of a single subrectangular room that must have been built in a single construction episode. The BLM stabilization crew (BLM 1975) thought a second story may have existed, but no evidence of it remains.

Tree-Ring Samples and Dates. Seven samples were collected from the site; four of the samples dated (Appendix). Samples from all six of the roof primary beams were collected, as was a single sample of roof closing material. Five of the seven samples are juniper; closing materials were exclusively juniper. All of the visible beam ends and interior limb stubs have been cut with a metal ax.

Interpretations. Simon Canyon pueblito presents an interesting interpretive situation. The evidence of metal-ax use indicates a postcontact occupation. The dates are all cutting or near-cutting dates, but they only weakly cluster. The structure must have been built in or after A.D. 1754. Thus, two of the roof primary beams (DNT-4 and DNT-5) were either reused or dead-wood elements. The absence of other structures in the area suggests that those beams represent the use of dead wood. The terminal rings indicate late spring or summer tree harvesting, and the structure was probably built in the spring or summer of A.D. 1754.

Rabbit House (LA 99804)

Rabbit House is a three-room pueblito located on the north side of the San Rafael valley floor across from Kin Tsézéí (Figure 4). It is one of the few pueblitos located on a valley floor, but it is on top of a boulder. It is clearly visible from Old Fort Ruin and Kin Tsézéí, and possibly several other sites as well, but Three Corn and Hidden House are not visible (Towner and Johnson 1998). A panel of ceremonial imagery located west of the site on a large sandstone boulder is probably associated with the occupation of the site.

Material Culture. The Rabbit House ceramic assemblage is sparse, but is dominated by Dinétah Gray and a few Gobernador Polychrome sherds. A few flaked stone artifacts are also present, but a detailed artifact inventory is unavailable.

Architecture. The architecture of the site suggests it was built in two distinct episodes that may not have been separated by a great length of time. Room 1 was the first room constructed, and Room 3 was probably built at the same time. Room 2 was abutted onto Room 3 at some later date.

Tree-Ring Samples and Dates. Twelve tree-ring samples have been collected, nine from in situ beams in the boulder-top rooms and three from loose logs around the base of the boulder (Appendix). Unfortunately the only dated in situ beams are from Room 1. Juniper was used exclusively in the construction, a somewhat atypical, but not unique, occurrence in pueblito sites. All but one of the beams show evidence of having been procured or modified with a metal ax.

Interpretations. The dating of Rabbit House indicates it was constructed during or after A.D. 1752, probably as a single construction episode. It

Figure 20. Santos Peak/Los Peñoles (LA 8948)

appears likely that dead wood was used for both roof primary and secondary beams. The dates, in conjunction with the intervisibility of Rabbit House and other pueblitos in San Rafael Canyon, suggest that it was contemporary with Old Fort Ruin and part of the Old Fort community (Towner and Johnson 1998).

Los Peñoles (LA 8948)

Los Peñoles is located on Santos Peak (Figure 20) in the northeastern part of the Dinétah (Figure 4). It has not been fully recorded, and the only date available at this time is a single tree-ring date of 1695vv derived from a pinyon ladder (presumably a notched-log ladder) on the north slope of the site (Appendix). The site was not included in the line-of-sight study (Jacobson et al. 1992), but must be visible from several sites in the northern part of Dinétah.

Hendricks and Wilson (1996:28–30) indicate that this site was the location of a fierce battle between Roque de Madrid's forces and the Navajos, led by Perlaja, on August 12, 1705. The inaccessibility of the site and tenacity of the Navajo defense prevented the Spaniards from ascending the peak. Thus, the Navajos gained a tactical victory. The single tree-ring date does nothing to contradict the history.

ROCKSHELTER SITES

Six sites consist of masonry architecture inside rockshelters or overhangs. Four sites are associated with forked-pole hogans on adjacent mesa tops, and surveys near the others have yet to be conducted. The masonry of the rockshelter sites is predominately jacal, but coursed masonry of unshaped stones is also present. All rockshelter sites are located in the central Largo area (Figure 4) and all of their tree-ring data are presented in Table 10. Only the Buffalo Mask Site and Morris Cliff Dwelling are discussed here; the other rockshelter sites are discussed in Towner (1997).

Buffalo Mask Site (LA 99809)

The Buffalo Mask site is a small two- or three-room structure located in an alcove high above a tributary of Delgadito Canyon. No hogans are associated with the site, but a portion of a corral fence is located on one of the benches below the site. The site is also loosely associated with some ceremonial imagery in the same canyon (Copeland and Rogers 1996). Buffalo Mask has a limited view down the tributary canyon and is one of the most inaccessible sites in the Dinétah; it is not visible from any other pueblito.

Material Culture. A few Dinétah Gray sherds were the only artifacts seen on the site.

Architecture. Room 1 is the only room with standing walls. Room 2 may simply be a storage area resulting from the extension of the south wall of Room 1. Room 3 may include more than one room, but only a small portion of the south wall remains. The rest of the rooms in the alcove have collapsed into a rubble pile.

Tree-Ring Samples and Dates. Sixteen tree-ring samples were collected from the site, but one was lost prior to accession into the LTRR system. Twelve of the remaining samples were collected from the alcove area and three were collected from the corral (Table 10). The samples from the alcove yielded eleven dates, including four cutting, two near-cutting, and five noncutting dates. The corral yielded two cutting dates. The beams in the alcove consist of eight juniper and four pinyon specimens. The species distribution contrasts with that at Shándin Bíikin, a very similar site

Table 10. Tree-ring Samples Collected from All Rockshelter Sites

Site Name Number	Sample Number	Species	Provenience	Inside Date	Outside Date	Terminal Ring	Tool Marks	Comment
BUFFALO	DNT-623	JUN	Loose log	1613	1680G	inc	MA limb, end	
MASK	DNT-624	JUN	Loose log	1606	1676G	inc	MA limb, broken end	
LA 99809	DNT-626	JUN	Loose log	1593	1672+vv	inc	MA limb, end	
	DNT-625	JUN	Horizontal intramural	1529	1670+G	inc	MA limb, broken end	
	DNT-627	PNN	Horizontal intramural	1575	1679+G	inc	MA limb	fork
	DNT-628	JUN	Horizontal intramural	1536	1670++B	inc	MA limb	fork
	DNT-629	JUN	Loose log RM 1	1535	1680GB	inc	broken limb, ends	
	DNT-630	PNN	Roof primary RM 1	1601	1673G	inc	MA limb	
	DNT-631	JUN	Loose log RM 1	NO DATE	—		MA end	
	DNT-631b		Loose log RM 1	LOST	—		broken ends	
	DNT-632	JUN	Loose log RM 1	1492	1675+vv	inc	broken limb, end	
	DNT-633	PNN	Loose log RM 1	1613	1678vv	inc	MA limb	
	DNT-634	PNN	Ladder Pole	1589	1692G	inc	MA end, broken limb	
	DNT-635	JUN	Corral	NO DATE	—		MA end	
	DNT-636	PNN	Corral	1841p	1915G		MA limb, end	
	DNT-637	PNN	Corral	1845p	1918G	inc	MA limb, ends	
LONE	DNT-458	PNN	Door 1 lintel	1550p	1652+LGB		broken ends	
WARRIOR	DNT-459	PNN	Door 1 lintel	1564p	1682++B		broken limb, end	
PUEBLITO	DNT-460	PNN	Door 1 lintel	1565	1674++vv	inc	broken limb, end	same as DNT-461
LA 99802	DNT-461	PNN	Door 1 lintel	—	—		broken ends	same as DNT-460
	DNT-462	JUN	Door 2 lintel	1455+/-	1646++vv		broken limb, ends	
	DNT-463	PNN	Door 2 lintel	1553	1627++vv		broken ends	
MORRIS	DNT-696	JUN	Room 7 Door lintel	NO DATE	—		MA end, broken limb	
CLIFF	DNT-697	PNN	Room 7 Door lintel	NO DATE	—		broken end	
DWELLING	DNT-698	PNN	Room 5 Loose, Jacal?	1530	1691++B		MA limb, ends	
	DNT-699	JUN	Room 8 Loose, Jacal?	1581	1675B	inc	broken ends	
	DNT-704	JUN	Room 8, Jacal?	NO DATE	—		broken ends	

	DNT-700	PNN	Room 8, Jacal	1656	1720B	comp	MA limb, ends	
	DNT-701	JUN	Room 1 Jacal	NO DATE	—		broken end	
	DNT-705	PNN	Room 1 Jacal	NO DATE	—		socketed, no marks	
	DNT-706	JUN	Room 1 Jacal	NO DATE	—		socketed, no marks	
	DNT-707	PNN	Room 1 Jacal	NO DATE	—		socketed, no marks	
	DNT-702	PNN	Room 4, loose	1670	1717vv	inc	MA ends	
	DNT-703	JUN	Room 4, loose	NO DATE	—		broken ends	
	DNT-709	PNN	Room 4, jacal	1622p	1720B	comp	socketed, no marks	
	DNT-708	JUN	Room 2, Jacal	NO DATE	—		socketed, no marks	
	DNT-710	JUN	Hogan leaner	NO DATE	—		weathered	
	DNT-711	JUN	Hogan leaner	1421	1695vv	inc	weathered	
	DNT-712	JUN	Hogan leaner	NO DATE	—		weathered	
	DNT-713	JUN	Ax-cut Stump	1509+/-	1726vv	inc	weathered	
	DNT-714	JUN	Hogan leaner	1455+/-p	1667++vv	inc	weathered	
	DNT-715	JUN	Hogan leaner	NO DATE	—		weathered	
RINCON	DNT-691	PNN	Door lintel	1666	1713L	comp		
ROCKSHELTER	DNT-692	PNN	Door lintel	1635p	1713L	inc		
LA 55835	DNT-693	PNN	Door lintel	1669	1713L		MA end	
	DNT-694	PNN	Split Door lintel	1668	1713L		MA limb	same as DNT-695
	DNT-695	PNN	Split Door lintel	—	—			same as DNT-694
SHAFT	DNT-312	JUN	Loose log in Upper Room	NO DATE	—		MA limb, ends	
HOUSE	DNT-313	JUN	Loose log in Upper Room	1528+/-	1667+vv	inc	MA limb	
LA 5660	DNT-314	JUN	Loose log in Upper Room	1493+/-	1694vv	inc	none	
	DNT-315	JUN	Loose log in Upper Room	NO DATE	—		MA limb, ends	
	DNT-316	JUN	Loose log in Upper Room	1554	1693vv	inc	MA limb	
	DNT-317	JUN	Loose log in Upper Room	NO DATE	—		MA end	fork
	DNT-318	JUN	Loose log in Upper Room	NO DATE	—		MA limb	
	NMS-8	JUN	Loose log in Upper Room	NO DATE	—			
	DNT-319	JUN	Upper Room Door lintel	1557	1712+LB	comp	MA shape	flat
	DNT-100	POP	Loose log on slope	NO DATE	—		none	
	NMS-1	JUN	House 3 (?)	NO DATE	—			

Table 10 (Cont.). Tree-ring Samples Collected from All Rockshelter Sites

Site Name Number	Sample Number	Species	Provenience	Inside Date	Outside Date	Terminal Ring	Tool Marks	Comment
SHÁNDIN	DNT-677	PNN	Roof primary	NO DATE	—		MA ends, limb	
BÍIKIN	DNT-678	JUN	Roof primary	NO DATE	—		MA ends, limb	
LA 106572	DNT-679	POP	Roof primary	NO DATE	—		MA ends, limb	
	DNT-680	POP	Roof primary	NO DATE	—		broken limb, ends	
	DNT-681	POP	Roof primary	NO DATE	—		broken limbs, ends	
	DNT-687	PNN	Roof primary	NO DATE	—		MA limb	
	DNT-683	JUN	Roof Support	NO DATE	—		MA ends, limb	
	DNT-684	PNN	Roof Support	NO DATE	—		MA limb, end	fork
	DNT-685	POP	Roof secondary	NO DATE	—		MA end, limb	
	DNT-686	JUN	Roof secondary	1584	1653vv	inc	none	
	DNT-682	PNN	Loose log	NO DATE	—		none	
	DNT-688	PNN	Loose log	1637p	1705GB		MA ends, limb	
	DNT-689	PNN	Ladder	1645	1702vv	inc	MA ends, limb	
	DNT-690	PNN	Ladder	NO DATE	—		MA ends, limb	

Key:
DNT= Tree-ring Lab Collection
NMS= Douglass Collection
PNN= Pinyon
JUN= Juniper
POP= Cottonwood
MA= Metal ax
limb= limb trim
inc= incomplete
comp= complete

setting, where pinyon and cottonwood both were used more frequently than juniper. Ten of the beams show evidence of procurement or modification with a metal ax, and six exhibit broken limbs.

Interpretations. The dates from the corral clearly indicate an early twentieth-century use of the area, probably by Hispanic sheep herders, and are consistent with dates from all other sampled corrals associated with pueblitos. The dates from the alcove, although somewhat scattered, suggest a single construction event in A.D. 1680. Both the Room 3 area and Room 1 contain beams that were cut during the 1680 growing season, as well as beams that date slightly earlier. There is some indication of dead-wood use (DNT-628), but other beams (DNT-634, 630) suggest reuse from an earlier structure. The terminal rings on all the beams with true outsides indicate wood procurement during the growing season, probably in the late spring or early summer. The climbing pole below the alcove yielded a 1692 cutting date that indicates the site was in use more than a decade after the masonry structure was built. Given the hiatus in dates and very sparse trash at the site, however, it is unlikely the site was ever occupied as a habitation; it was probably used only on a temporary basis. Nevertheless, Buffalo Mask is the earliest securely dated pueblito in the Dinétah.

Morris Cliff Dwelling (DCA-88-118)

Morris Cliff Dwelling is an eight-room structure in an alcove at the head of Adams Canyon. The site includes at least five forked-pole hogans on the mesa above the alcove. No other sites can be seen from either the rockshelter or the hogan area. The site was visited by Morris (Carlson 1965:44), who made a pottery collection and excavated some burials in the vicinity.

Material Culture. Morris lists only Gobernador Polychrome, Puname Polychrome, Tewa Polychrome, and Pueblo I intrusives from the site (Carlson 1965:45). Hancock and Johnson (1988) report Dinétah Gray, Gobernador Polychrome, and Jemez Black-on-white sherds from the portion of the site above the alcove; no data are available from the alcove itself.

Architecture. Currently, none of the rooms in the alcove are attached to any of the other rooms. Pecked post holes that functioned as sockets for jacal wall beams delineate the architectural layout. These sockets suggest that rooms 1 through 5 and Room 8 were built as a unit. Rooms 6 and 7 appear to have been unattached units. None of the hogans on top of the mesa can be architecturally linked to the others or to construction in the alcove.

Tree-Ring Samples and Dates. A total of 21 samples has been collected from the site, 15 from the alcove rooms, five from the hogans, and a single sample from an ax-cut stump associated with the hogans (Table 10). The 15 samples from the alcove rooms yielded five dates—three cutting and two noncutting; the hogan and stump samples (n=6) yielded three noncutting dates. All of the hogan samples, including the stump, are juniper; the alcove produced eight pinyon and seven juniper samples. Four of the beams in the alcove show evidence of harvesting and/or modification with a metal ax; six of the samples have broken ends or limbs, often indicative of dead-wood use. Sample DNT-699 has broken ends and dates anomalously early at 1675B; it is clearly a dead-wood element, as it DNT-698 which dates 1691++B. All of the hogan samples were too eroded for tool marks to be identified.

Interpretations. The two 1720 cutting dates from Rooms 8 and 4 suggest that the entire structure was built in a single construction episode in that year. Both these pinyon samples (DNT-700, 709) exhibit complete terminal rings, indicating that they were cut after the end of the pinyon growing season, probably in the fall of A.D. 1720 or winter of 1721. Although two dates are not conclusive, I believe all of the rooms in the alcove were built in the fall or winter of 1720–1721. This interpretation is not contradicted by the noncutting 1717vv date from Room 4. The samples from the hogans, however, provide little temporal information other than that they were occupied sometime after the seventeenth century. The sample from the ax-cut stump (DNT-713) probably dates the occupation to some time in the 1720s and suggests that part of the site was occupied, at least intermittently, for six years.

MESA RIM PUEBLITOS

The largest group of sites in the analysis are mesa rim pueblitos, which consist of masonry structures on, or immediately adjacent to, sandstone mesa rims or benches. These sites typically have expansive viewsheds in

Table 11. Mesa-rim Pueblitos Used in the Analysis

Discussed	Not Discussed	
Frances Canyon Ruin	Adams Canyon Pueblito	Tower of the Standing God
Garcia Canyon Pueblito	Hummingbird House	Cabresto Mesa Tower
Kin Wólá Jíní	Hadlock's Crow Canyon Pueblito	Kin Tsézéí
The Overlook Site	Largo School Pueblito	LA 2297
Compressor Station	Muñoz Canyon Pueblito	Pablo Spring Site
Gould Pass Pueblito	Pueblito Canyon Pueblito	
Hooded Fireplace Ruin	Pueblito East	

one direction, but relatively unimpeded access from another. Most mesa rim sites are associated with forked-pole hogans and other features, but many have not been included in intensive areal surveys (cf. Marshall 1991; Wilshusen et al. 2000). Mesa rim pueblitos are present throughout the Dinétah (Figure 4; Table 11). Only six examples are discussed here, but the others are discussed in Towner (1997).

Frances Canyon Ruin (LA 2135)

Frances Canyon Ruin is located in a small side canyon west of the main branch of Frances Canyon (Figure 4). It is one of the largest, most complex pueblito sites in the Dinétah and contains at least 40 rooms, including second- and third-story rooms and a four-story tower complex (Figure 21). Intramural features include storage bins, wall pegs, floor hearths, loom holes, and two hooded fireplaces. Four sweatlodge locations were identified near the pueblito, but none contain any superstructure remains. At least eight middens are located near the site. A single forked-pole hogan (LA71596) is present approximately 300 m southeast of the pueblito, and two masonry-based hogans, two hearth areas, and a pen or corral enclosure are present as well (Marshall 1991, 1995). Jacobson et al. (1992) indicate that the Cabresto Mesa Tower Complex (LA 2138) and Crevice Ruin (LA 13218) can be seen from Frances Canyon Ruin, but the line-of-site to Cabresto Mesa tower has not been field verified.

Material Culture. A wide variety of items from excavated (Carlson 1965; Keur 1944) and surface contexts (Powers and Johnson 1987; Marshall 1991, 1995) has been documented at the site. The ceramic assemblage consists predominately of Dinétah Gray and Gobernador Polychrome sherds, although small numbers of Zia-Puname, Tewa, Acoma, and Jemez

Figure 21. Frances Canyon Ruin (LA 2135)

wares are also present (Marshall 1995:24). The abundant flaked and ground stone tools are made of a variety of materials including Pedernal Chert, obsidian, Washington (Narbona) Pass Chert, and local sandstones and igneous rocks. Faunal remains include those of a duck-sized bird, horse/burro, domestic sheep, domestic sheep/goat, and deer. The site contains the largest number of historical artifacts documented at any pueblito, including metal knife parts, gun parts, an iron ax, a metal cross, three copper bells, an awl, a needle, a nail, a ring, and abundant glass trade beads (Carlson 1965; Marshall 1995).

Architecture. Because wood samples are derived only from a portion of the rooms at the site (see below), wall analysis has an important part in the research on construction sequences at Frances Canyon. Twenty-five architectural units (A.U.) can be defined (Figure 22), based on Morris's original map. Few of these relationships can be discerned today, so Morris's map provides most of the architectural data. Five of these units, including Morris's hogan (A.U. 22), Room 40 (A.U. 21), Marshall's (1995:18) two possible forked-pole hogans (A.U. 23 and 24), and the "enclosure" (A.U. 25), cannot be related to the rest of the structures on

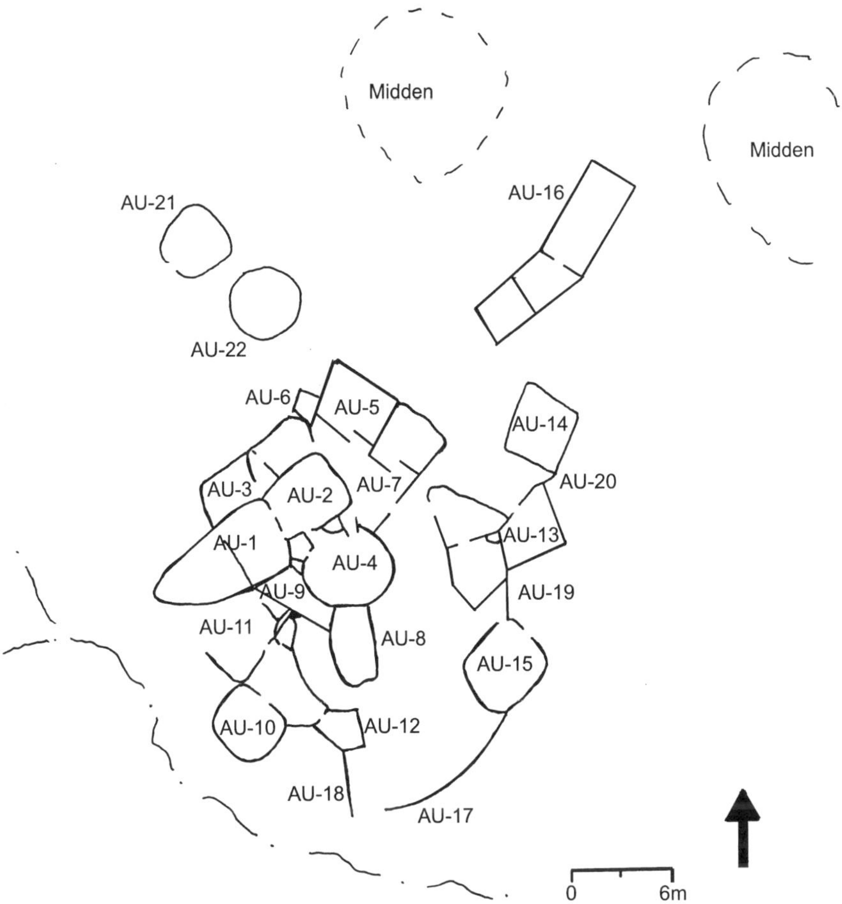

Figure 22. Architectural units defined at Frances Canyon Ruin (LA 2135). (After Carlson 1965, based on map by Earl Morris)

architectural criteria alone. Several other architectural units are more integral to the pueblito, but also are free-standing units.

Architectural Unit 1 comprises the "tower" for which the site is famous. The tower itself originally consisted of seven rooms. The lowest level contains Room 3 and an unnumbered room, both of which were small storage rooms; the second level includes habitation Rooms 1 and 4; the third story consists of Rooms 2 and 5; and the fourth story, which no longer exists, contained Room 6 (Carlson 1965:32–33). Obviously, the upper rooms must have been built after the lower rooms, and the masonry suggests most of the tower was built in a single construction episode.

Architectural Unit 2 consists of Room 10, which was abutted to the boulder below the tower and was connected to Room 4 by a doorway; Rooms 11 and 12 comprised the second and third levels above Room 10 and may have been connected to Rooms 5 and 6, but wall deterioration prevented Morris or Carlson from making such an inference. These rooms must have been built after the tower was completed.

Architectural Unit 3 includes Rooms 7, 8, and 9. Room 8 was abutted to Room 10; Room 9 comprised the second story above Room 8; Room 7 was created by an interior wall separating it from Room 8. Thus, Rooms 7 through 9 must have been built later than Room 10 (A.U. 2) and Room 4 (A.U. 1).

Architectural Unit 4 consists only of Room 18 and the storage bins in it (see Carlson 1965:39, Plate 13). Room 18 was abutted to Room 10 (A.U. 3) and connected to it by a doorway.

Rooms 14–17 constitute Architectural Unit 5. Little remains of these rooms today, but Morris indicates that a two-story structure existed in 1915. Because the west wall of Room 14 abuts the northeast wall of Room 8, A.U. 5 must have been built later than A.U. 3.

Architectural Unit 6 includes only Room 13, a small storage room abutted to Room 14. Therefore, it must postdate Room 14.

Architectural Unit 7 comprises only the courtyard identified by Morris. The only construction associated with this courtyard is the building of a wall between Room 16 (A.U. 5) and Room 18 (A.U. 4). This wall appears to be a continuation of the southeast wall of Room 16 and also appears to be abutted to the northeast wall of Room 18. This suggests that the courtyard is contemporaneous with A.U. 5 construction, but postdates A.U. 4 construction.

Architectural Unit 8 includes Room 19 and its second story counterpart, Room 20. These rooms are abutted to Room 18 and a doorway originally led from Room 19 onto the roof of Room 18 (Carlson 1965:34).

Architectural Unit 9 consists of Room 21. It is an unroofed area between the tower boulder and Rooms 18 and 19. The southern wall of Room 21 is the only wall built as part of this room and was abutted to the west wall of Room 19.

Architectural Unit 10 contains only Room 25. It appears as if this, the southernmost room at the pueblito, was originally a free-standing structure. The walls that constitute A.U. 11, however, are abutted to Room 25 and must postdate its construction.

Rooms 22, 23, and 24 constitute Architectural Unit 11. These rooms cannot be related temporally to Architectural Units 1–9 on the bases of architectural relationships because they are slightly lower and were built between boulders. The southeast wall of Room 23 is abutted to the northwest wall of Room 25, however, and indicates that A.U. 11 postdates A.U. 10.

Architectural Unit 12 includes only Room 26 which abuts Room 23. Therefore, it must postdate A.U. 11.

Architectural Unit 13 comprises Rooms 29–33. The map is somewhat ambiguous, but it appears as if Rooms 29–31 were built as a unit; the north wall of Room 31 abuts Room 32 and the south wall of Room 32 abuts Room 29, however. Thus, if the Morris map is correct, all five rooms must have been built at the same time.

Architectural Unit 14 contains Room 34 and its second story, Room 35. Apparently, this was a free-standing structure, and it cannot be associated with the other structures on an architectural basis.

Rooms 27 and 28 (A.U. 15) comprise a free-standing two-story structure on a rock outcrop southeast of the tower. They cannot be incorporated into an architecturally based construction sequence because they contact no other rooms.

Architectural Unit 16, a free-standing room-block on the north end of the site, includes Rooms 36–39. These rooms must have been built as a unit, but because they contact no other structures, they cannot be incorporated into an architecturally based construction sequence.

Architectural Unit 17 is the wall that extends southwest from Room 27; it abuts Room 27 and must have been built after that room was complete. Architectural Unit 18 is the wall that extends south from Room 26 (A.U. 12). It must have been built later than that room because it abuts the southern wall. Architectural Unit 19 is the wall that connects Rooms 27/28 (A.U. 15) and Rooms 29–33 (A.U. 13). As such, it must postdate those architectural units. Similarly, Architectural Unit 20 consists of the small wall segment that connects Architectural Units 13 and 14; it must postdate both of those architectural units.

Room 40, a free-standing room north of the main room-block, constitutes Architectural Unit 21; it cannot be architecturally associated with any of the other structural remains. Finally, Architectural Unit 22 is the hogan between Room 40 and the main room-block; likewise, it cannot be architecturally associated with any of the other units.

Tree-Ring Samples and Dates. A total of 53 samples has been collected from Frances Canyon Ruin, including one duplicate (Figure 23; Table 12). Earl Morris collected a single sample (H-6), the Laboratory of Anthropology collected six samples (RG-1–6); the Navajo Land Claim personnel collected six (NLC-983–986, 1055); and this project collected 40 samples (DNT-8–47). Discounting the one duplicate sample, the 52 independent samples yielded 48 dates, including 33 cutting, 6 near-cutting, and 9 noncutting dates. The samples that did not date include DNT-11 from a ponderosa pine roof beam in Room 18, a minimally provenienced ponderosa pine sample (NLC-984) that may be from the tower complex, a pinyon sample from the same area (NLC-983), and a pinyon sample from a hogan (NLC-982).

The species distribution at Frances Canyon is unusual among the pueblito sites. Although pinyon (n=23) and juniper (n=16) are the dominant species, the relatively high number of Douglas-fir specimens (n=10) and presence of ponderosa pine (n=3) are somewhat unusual. The lack of *Populus* sp. in the samples is probably a reflection of dendrochronological sampling bias; the hooded fireplace support in Room 29 appears to be cottonwood and was not sampled. It is interesting to note that different species were exploited at different times. The juniper samples date primarily to the early 1710s, although two beams were cut in 1722; the pinyon samples date in the mid- to late-1710s, and harvesting increased in the early 1720s; the single dated ponderosa pine was harvested sometime after 1728; finally, Douglas-fir trees were exploited only during the late 1730s and early 1740s.

Similarly, tree-harvesting seasons, as indicated by terminal ring analysis, suggest that Frances Canyon is different from other sites. The majority of the pinyon and juniper specimens, particularly those used during the early construction episodes at the site, were harvested after the growing seasons ended, probably in the late fall or early spring. Most of the Douglas-fir specimens used in the later construction events at the site, however, were cut during the growing season, probably in March, April, or May.

All of the samples for which beam attributes have been recorded show signs of metal ax use. This is not surprising because Frances Canyon is one of the pueblito sites where Spanish-style metal axes have been found.

Construction Sequence. Frances Canyon Ruin shows different construction episodes very clearly. The building of Room 3 in A.D. 1710 was the first construction event at the site, as evidenced by the strong cluster of

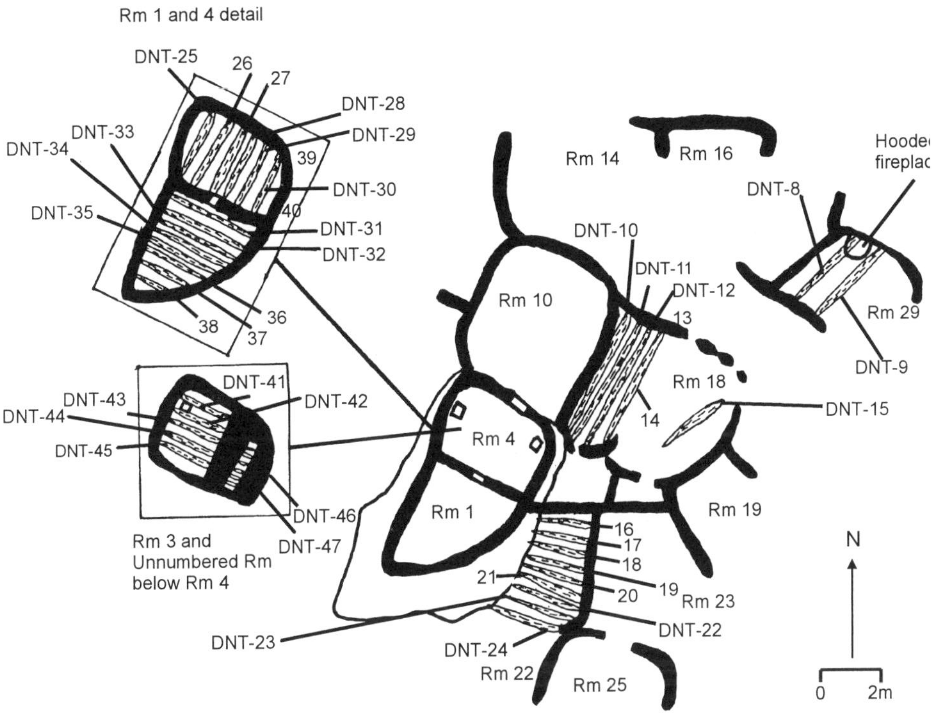

Figure 23. Tree-ring sample proveniences at Frances Canyon Ruin (LA 2135). (After Powers and Johnson 1987)

five 1710 cutting dates. All of the beams in this room are small, and the room is less than 1 m high and less than 2 m long. It was undoubtedly built as a storage facility. All of the samples from Room 3 have complete terminal rings, indicating construction after the 1710 juniper growing season ended. In or shortly after A.D. 1714, the unnumbered room was built adjacent to Room 3 as evidenced by the two 1714 dates (one cutting). Interestingly, the remainder of the tower (A.U. 1) was not built until 1722.

Rooms 1 and 4 were added to the tower in the fall of 1722 or winter of 1722–1723, after the 1722 pinyon and juniper growing seasons ended. Room 1 was constructed with predominately freshly cut timbers, although two beams (DNT-34, 35) may have been reused from another structure. These beams suggest another structure may have been associated with the 1714 construction episode, but was subsequently dismantled. Room 4, however, was built using freshly cut beams (DNT-26, 27, 29), dead-wood elements (DNT-25, 28), and a reused beam (DNT-30).

Table 12. Tree-ring Samples Collected from Frances Canyon Ruin (LA 2135)

Sample Number	Species	Provenience	Inside Date	Outside Date	Terminal Ring	Tool Marks	Comment
DNT-8	DF	Roof primary RM 29	1685	1743G	inc	MA bark, limb	SB
DNT-9	DF	Roof primary RM 29	1693	1743vv	inc	MA bark, limb	
DNT-10	PP	Roof primary RM 18	1677	1728vv	inc	MA bark, limb	SB
DNT-11	PP	Roof primary RM 18	NO DATE			MA bark, limb	SB
DNT-12	DF	Roof primary RM 18	1680	1736GB	comp	MA bark, limb	charred?
DNT-13	DF	Roof primary RM 18	1658	1736v	comp	MA bark, limb	SB
DNT-14	DF	Roof primary RM 18	1611	1733vv	inc	MA bark, limb, shape	flat; no SB
DNT-15	PNN	Loose log RM 18	1410	1721vv	inc	MA bark, limb, end	
DNT-16	JUN	Roof primary RM 22	1617	1714+B	comp	MA limb	SB
DNT-17	PNN	Roof primary RM 22	1575	1718+B	comp	MA limb	SB
DNT-18	PNN	Roof primary RM 22	1543	1715+B		MA limb, shape	flat; SB
DNT-19	JUN	Roof primary RM 22	1537	1716B	comp	MA limb, shape	flat; SB
DNT-20	JUN	Roof primary RM 22	1654	1714+B	comp	MA limb, shape	flat; SB
DNT-21	PNN	Roof primary RM 22	1488	1720vv	inc	MA limb	SB
DNT-22	PNN	Roof primary RM 22	1607	1715B	inc	MA bark, limb, shape	flat; SB
DNT-23	JUN	Roof primary RM 22	1425	1715B	inc	MA limb	no SB
DNT-24	PNN	Roof primary RM 22	1616p	1716vv	inc	MA limb	no SB
DNT-25	JUN	Roof primary RM 4	1436+/-	1689++B		MA limb	SB
DNT-26	PNN	Roof primary RM 4	1564	1722v	comp	MA limb, shape	flat; SB
DNT-27	PNN	Roof primary RM 4	1425	1722v	comp	MA limb, shape	flat; SB
DNT-28	JUN	Roof primary RM 4	1384	1672B	inc	MA limb, shape	flat; SB
DNT-29	PNN	Roof primary RM 4	1572	1722v	comp	MA limb, shape	flat; SB
DNT-30	PNN	Roof primary RM 4	1536p	1715+B	comp	MA limb, shape	flat; SB
DNT-39	PNN	Smoke hole RM 4	1674	1717G	inc	MA limb	SB
DNT-40	PNN	Smoke hole RM 4	1678p	1717G	comp		SB
DNT-31	JUN	Roof primary RM 1	1631+/-	1722v	comp	MA limb, shape	flat; SB
DNT-32	PNN	Roof primary RM 1	1584p	1722v	comp	MA limb, shape	flat; SB
DNT-33	JUN	Roof primary RM 1	1491	1722v	comp	MA bark, limb	SB
DNT-34	JUN	Roof primary RM 1	1606+/-	1711v	comp	MA bark, limb	SB
DNT-35	JUN	Roof primary RM 1	1539+p	1713+B		MA bark, limb, shape	flat; SB; reused
DNT-36	PNN	Roof primary RM 1	1576	1722v	comp	MA limb	SB
DNT-37	PNN	Roof primary RM 1	1568	1722v	comp	MA limb	SB
DNT-38	PNN	Roof primary RM 1	1587	1722B	comp	MA limb	SB
DNT-41	JUN	Roof primary RM 3	1504	1710B	comp	MA limb	SB
DNT-42	JUN	Roof primary RM 3	1560+/-	1710B	comp	MA limb	SB
DNT-43	JUN	Roof primary RM 3	1526p	1710v	comp	MA limb	SB
DNT-44	JUN	Roof primary RM 3	1284+/-	1710B	comp		split; SB
DNT-45	JUN	Roof primary RM 3	1467	1710B	comp	MA limb	SB
DNT-46	PNN	Roof secondary un RM	1662	1714vv	inc	MA limb	SB
DNT-47	JUN	Roof secondary un RM	1653	1714v	inc	MA limb	SB
H-6	DF	Unknown	1655p	1735v	inc	Unknown	
RG-789-1	DF	Unknown	1648p	1737v	inc	Unknown	
RG-789-2	—	Unknown	—	—	—	Unknown	same as DNT-8
RG-789-3	PNN	Unknown	1603p	1721v	inc	Unknown	
RG-789-4	PNN	Unknown	1614p	1716v	inc	Unknown	

Table 12 (Cont.). Tree-ring Samples Collected from Frances Canyon Ruin (LA 2135)

Sample Number	Species	Provenience	Inside Date	Outside Date	Terminal Ring	Tool Marks	Comment
RG-789-5	PNN	Unknown	1540p	1717+vv	inc	Unknown	
RG-789-6	DF	Unknown	1698p	1743c	inc	Unknown	
NLC-982	PNN	Hogan 2	NO DATE			Unknown	
NLC-983	PNN	Unit 1 Tower	NO DATE			Unknown	
NLC-984	PP	Unit 1 Tower primary	NO DATE			Unknown	
NLC-985	PNN	Unit 1 Tower	1677p	1745c	inc	Unknown	
NLC-986	DF	Fortified Crag 1	1634p	1743incG	inc	Unknown	
NLC-1055	DF	House Unit 1	1697p	1743inc	inc	Unknown	

Key: RG= Rio Grande Collection; NLC= Navajo Land Claim Collection; H= Douglass Collection; DNT= Tree-ring Lab Collection; PNN= Pinyon; JUN= Juniper; DF= Douglas-fir; PP= Ponderosa pine; MA= Metal ax; limb= limb trim; bark= debarking; shape= shaping; inc= incomplete; comp= complete; SB= Smoke blackening present

Interestingly, the closing material above the roof primaries (DNT-39,40) predates the construction of the room and must have been reused from another structure. If sample RG-789-3 does indeed come from the "top of the tower," as indicated by Hall (1951), it may have been cut for use in Rooms 1 or 4, but not used until Architectural Unit 1 (the tower) was completed after the 1722 growing season ended.

The only other early (pre-1730) tree-ring-dated construction episode was the building of Room 22. Based on the tree-ring dates alone, there are two possible interpretations of the dates. First, the room may have been built in the late 1710s, between 1715 and 1718, but the mixture of near-cutting dates and lack of clustering makes identification of a specific year difficult. This room may then have been repaired sometime after A.D. 1720, probably during the construction of Rooms 1 and 4 in A.D. 1722. The second interpretation of the dates is that the entire room and architectural unit were built after 1720 with predominately reused beams. Architectural data support the latter interpretation (see below).

The next tree-ring-dated construction event was the building of Room 18 in A.D. 1736, probably after the growing season had ended. This room is one of the key architectural units on the site and was apparently roofed with freshly cut Douglas-fir beams; samples H-6 collected by Morris and RG-789-1 collected by Hall may also be from this area of the site.

Finally, the tree-ring dates indicate construction of Room 29 in the spring or summer of 1743, probably between March and the end of May, with freshly cut Douglas-fir timbers. Although there is no positive evidence, I suspect that sample NLC-1055 from "House Unit 1" also dates Room 29 construction, and NLC-986 may date it as well. I doubt, however, that NLC-985, which dates to 1745, was recovered from the "Unit 1 Tower." It is possible that the very top room of the tower, Room 6, was added in the 1740s, but because no other provenienced beams from that area of the site date later than 1722, I suspect the NLC sample is not from the tower.

Interpretations. The combination of the tree-ring and architectural data indicates a relatively complex picture of the occupation, growth, and abandonment of Frances Canyon Ruin. Clearly, Room 3 at the base of the tower is the earliest room built at the site (1710); four years later, in 1714, the unnumbered room was built next to Room 3 incorporating part of the exterior wall. The remainder of the tower was built after the end of the 1722 growing season.

The tree-ring dates suggest Architectural Unit 11 (Rooms 22–24) was built sometime in the years between 1714 and 1720. I believe that Architectural Unit 11 was built in 1722 at the same time as the tower was completed, although there is no solid evidence to support such a claim. Architectural Unit 11 does not abut any other structure and cannot be assigned a relative temporal position in the early construction episodes at the site; the only tree-ring samples in the unit are from Room 22, but they do not cluster well. The latest date from Room 22, 1720vv, is neither a cutting date nor repair beam and suggests construction after 1720. Construction of the architectural unit in 1722 is particularly appealing because Room 24, a second-story room, would have provided access to the tower across the roof of Room 22. It is possible, based on the near-cutting dates in the late 1710s, that other structures existed, but were torn down when the tower was constructed, their beams used to roof Room 22.

Architectural Unit 2 clearly postdates 1722; Architectural Unit 3 postdates A.U. 2, Architectural Unit 5 postdates A.U. 3, and Architectural Unit 6 postdates A.U. 5. Thus, it appears that by the winter of 1722 the site consisted of the completed tower, at least one, and possibly three suites of rooms to the south (A.U. 11 and possibly A.U.s 10 and 12), and three sets of rooms to the north and northeast (A.U.s 2, 3, and 5). The bins and doorway noted by Carlson properly belong to this architectural unit. This entire section of the site, however, predates Architectural Unit 4 (Room 18).

Room 18 (A.U. 4), the next construction at the site, was not built until fourteen years later in 1736. Architectural Unit 7, the courtyard, apparently was created either at the same time or shortly thereafter. Finally, Architectural Unit 8 (Rooms 19 and 20) postdates Room 18 and, therefore, 1736. Thus, after the 1736 Douglas-fir growing season, the site consisted of the tower and all the room-blocks that form the core of the complex. If Morris's sample, H-6, is from Room 18, it provides additional support for a mid-1736 construction episode.

The next documented construction event at the site, in the spring of 1743, consisted of Architectural Unit 13 (Rooms 29–33). The only dated room in this room-block, Room 29, contains both a cutting and a noncutting 1743 date; if samples NLC-986 and 1055 are from this same area, they provide additional support for a spring 1743 construction episode. It is probable that Architectural Units 15 (Rooms 27, 28) and 14 (Rooms 24, 25) were built at the same time, but there is no evidence to confirm or refute this suggestion. The walls connecting these architectural units to A.U. 13, however, suggest all the architectural units were occupied contemporaneously. Finally, if the two walls (A.U.s 17, 19) at the southern end of the site were connected—and I suspect that they were—then this "retaining" wall may have been one of the latest construction events at the site.

The lack of tree-ring dates from the hogans and other free-standing structures prevents me from assigning them a place in the construction sequence. One nearby hogan, LA 71596 (Marshall 1991), yielded three tree-ring dates in the mid- to late-1500s, but all are noncutting dates that probably predate the use of the structure by many decades (Towner 1997).

Summary. Frances Canyon Ruin appears to have been occupied in two or possibly three temporally discrete stages. The initial occupation between 1710 and 1714 involved the construction of the lower storage rooms of the tower, but no obvious habitation areas were constructed. The second part of the occupation involved the construction of the tower and Architectural Units 2, 3, 5, and 6 in 1722. The distribution of cutting and near-cutting dates throughout the late 1710s suggests that other structures may have been present and that the site was not abandoned between masonry construction episodes. Indeed, additional structures may have been built during this interval, but subsequently dismantled and some of the wood reused in tower construction.

Marshall (1995) uses the range of tree-ring cutting dates (1710–1745) and adds 10 years to generate a 45-year-long occupation estimate for Frances Canyon Ruin. I believe the site was abandoned for a period of time, probably at least ten years, after the tower complex was completed.

167	2
—	
171	00000134445556678
172	1222222222
173	5667
174	33335

Figure 24. Stem-and-leaf plot of cutting and near-cutting dates from Frances Canyon Ruin (LA 2135)

The cutting and near-cutting dates indicate almost continuous tree harvesting from 1710 through 1722 (Figure 24); trees were cut in 10 of the 13 years spanned. After 1722, however, no trees were cut until 1735. Tree harvesting resumed when Room 18 and the adjacent architectural units were built in 1736 and continued until the site was permanently abandoned sometime after 1745. The six-year gap between the 1737 and 1743 cutting dates may indicate temporary abandonment of the site, but may also be a function of small sample size and insufficient sampling of appropriate site areas.

Garcia Canyon Pueblito (LA 36608)

Garcia Canyon Pueblito is a nine-room structure adjacent to a small cliff face overlooking the confluence of Frances Canyon and a small drainage (Figure 4). The site has limited visibility to the south, west, and east, but is open to the north; it does not have lines-of-sight to any other pueblitos (Jacobson et al. 1992).

Material Culture. Powers and Johnson (1987:80) identified Dinétah Gray and Gobernador Polychrome ceramics and a piece of Mexican Majolica ware at the site. Also present were green quartzite and white chert lithic artifacts.

Architecture. Several construction episodes are apparent at the site (Figure 25). The first included building Rooms 1, 2, and 3, and may have included Room 4 as well. There is some uncertainty about the inclusion of Room 4 because of the destruction of the junction of the west wall of Room 4 and south wall of Room 2/3. The curvature of the west wall of Room 4 and contiguous nature of the east wall that connects Rooms 2 and 4 suggest that Room 4 was built as part of the same construction

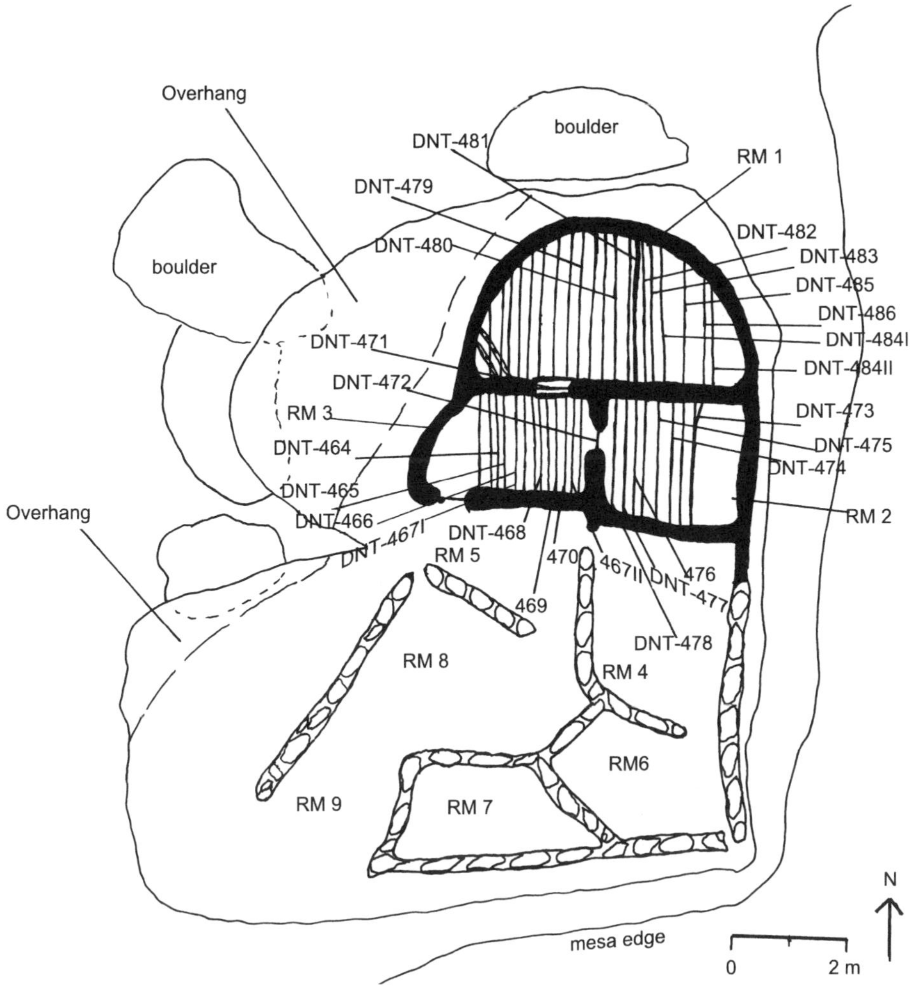

Figure 25. Tree-ring sample proveniences at Garcia Canyon Pueblito (LA 36608). (After Powers and Johnson 1987)

event. Rooms 6 and 7 were clearly abutted to Room 4 after it was complete. Rooms 8 and 9 are too deteriorated to evaluate in detail, but appear to have been abutted to Rooms 4 and 7. The last four rooms (6, 7, 8, and 9) show no evidence of ever having been roofed.

Tree-ring Samples and Dates. Twenty-seven tree-ring samples, all of which yielded dates, were collected from the site (Table 13). Seventeen cutting or probable cutting dates, one near-cutting date, and nine noncutting

Table 13. Tree-ring Samples Collected from Garcia Canyon Pueblito (LA 36608)

Sample Number	Species	Provenience	Inside Date	Outside Date	Terminal Ring	Tool Marks	Comment
DNT-464	JUN	RM 3 Roof primary	1658	1720LG	comp	MA limb, ends	
DNT-465	JUN	RM 3 Roof primary	1625+p	1717+vv	inc	MA limb, end	
DNT-466	PNN	RM 3 Roof primary	1638	1715G	inc	MA limb, ends	
DNT-467I	JUN	RM 3 Roof primary	1629	1699vv	inc	MA bark, limb	
DNT-467II	PNN	RM 3 Roof primary	1508	1713vv	inc	none	
DNT-468	JUN	RM 3 Roof primary	1610	1712L		MA limb, end	SB
DNT-469	JUN	RM 3 Roof primary	1604	1719G	comp	MA limb, end	SB
DNT-470	PNN	RM 3 Roof primary	1633	1712LGB	comp	MA bark, limb, end	SB
DNT-471	JUN	RM 1/3 Door lintel	1609	1701L	inc	MA limb	
DNT-472	PNN	RM 2/3 Door lintel	1620	1700vv	inc	Unknown	socketed
DNT-473	JUN	RM 2 Roof primary	1630	1719vv	inc	MA limb, ends	fork
DNT-474	JUN	RM 2 Roof primary	1613	1712G	inc	MA limb, ends	
DNT-475	PNN	RM 2 Roof primary	1641	1717+G	inc	MA limb	
DNT-476	PNN	RM 2 Roof primary	1610	1722G	inc	MA bark, limb, ends	
DNT-477	JUN	RM 2 Roof primary	1618	1712GB	inc	MA ends	
DNT-478	JUN	RM 2 Roof primary	1614	1722vv	inc	MA limb, end	
DNT-479	PNN	RM 1 Roof primary	1628	1714++vv		MA limb	
DNT-480	JUN	RM 1 Roof primary	1649	1720LGB	comp	MA end	
DNT-481	JUN	RM 1 Roof primary	1536+	1721LB	comp	MA limb, end	SB
DNT-482	PNN	RM 1 Roof primary	1644p	1722v	comp	MA limb, end	SB
DNT-483	JUN	RM 1 Roof primary	1603+p	1720LGB		MA limb, shape, end	flat; SB
DNT-484I	PNN	RM 1 Roof primary	1617p	1722L	comp	MA limb, end	SB
DNT-484II	PNN	RM 1 Roof primary	1607	1721LGB	comp	MA limb	SB
DNT-485	PNN	RM 1 Roof primary	1603	1721LB	comp	MA bark, limb	SB
DNT-486	PNN	RM 1 Roof primary	1600p	1722v	comp	MA bark, limb, end	SB
NLC-977	PNN	Unknown	1615p	1712+vv	inc	Unknown	
NLC-978	PNN	Unknown	1376p	1710+vv	inc	Unknown	

Key: DNT= Tree-ring Lab Collection; NLC= Navajo Land Claim Collection; PNN= Pinyon; JUN= Juniper; MA= Metal ax; limb= limb trim; bark= debarking; inc= incomplete; comp= complete; SB= Smoke blackening present

dates were derived from the samples and span a period of 23 years. Cutting-date clusters occur in 1712, 1720, 1721, and 1722. All of the samples, except the two NLC samples, were collected from in situ beams. The species distribution is evenly split between pinyon (n=14) and juniper (n=13). Because virtually all the samples are from in situ beams, the distribution is clearly a result of the inhabitants' selection criteria and not a result of dendrochronological bias. All of the beams, except a single roof primary in Room 3 and a socketed door lintel, show evidence of procurement or modification using a metal ax. The majority of beams in Room 1 and three of the beams in Room 3 are smoke blackened, an indication of hearth fires or other habitation-related activities.

Interpretations. The dendrochronological data from Garcia Canyon Pueblito indicate complex behaviors regarding the construction of the structure and use of wood as a resource. The cluster of 1712 cutting dates indicates a major tree-harvesting event at that time, but these dates are in both Room 2 and 3, which must have been built at the same time as Room 1. Room 1, however, contains cutting-date clusters in 1720, 1721, and 1722, but no dates earlier than 1714, a probable dead-wood element. Room 3 contains no beams dated later than 1720, yet must have been built at the same time as Rooms 1 and 2. Finally, Room 2 contains beams cut in 1712, 1717 (probably), and 1722. Door lintels sometimes offer clues to interpreting such architectural and dendrochronological problems, but the lintels at Garcia Canyon (DNT-471, 472) are clearly dead-wood elements that date much earlier than the construction of the building.

I interpret the architectural and dendrochronological data as follows: Rooms 1, 2, 3, and 4 were all constructed in 1712; minor repairs occurred in 1715, 1717, and possibly 1719. Anticipating major remodeling, the site occupants stockpiled beams in 1720 and 1721; these beams were used with freshly cut beams near the end of the 1722 growing season when the entire structure was reroofed. One dead-wood element, DNT-479, was used as part of the remodeling. I believe that Rooms 6–9 were also added in 1722, but there is no dendrochronological evidence to support that inference. The site must have been abandoned within a few years, certainly by 1730, or additional repairs would have been required.

Kin Wólá Jíní (LA 12285)

Kin Wólá Jíní was only minimally recorded in the early 1970s. It is a seven-room masonry structure on a mesa edge overlooking Cereza Canyon and is one of the easternmost of all known pueblitos (Figure 4). No other known sites are visible from the pueblito.

Material Culture. Field observations include Dinétah Gray and Gobernador Polychrome ceramics, flaked and ground stone artifacts, an associated sweatlodge, and corn-stalk petroglyphs on the cliff face below the structure.

Architecture. The masonry structure consists of seven rooms and was built in one or two construction episodes (Towner 1997:199; Figure 45). Rooms 3–7 were certainly built as a unit. Rooms 1 and 2 may have been

Table 14. Tree-ring Samples Collected from Kin Wólá Jíní (LA 12285)

Sample Number	Species	Provenience	Inside Date	Outside Date	Terminal Ring	Tool Marks	Comment
DNT-423	PNN	RM 6 Roof Auxilary	1458	1705+vv	inc	MA limb	
DNT-424	PNN	RM 6 Roof primary	1565	1711vv	inc	MA limb, end	
DNT-425	PNN	RM 6 Roof primary	1633p	1710vv	inc	MA limb	
DNT-426	PNN	RM 6 Roof primary	1433	1649++vv	inc	MA limb, end	
DNT-427	PNN	RM 6 Loose log	1465	1703+vv	inc	MA limb	
DNT-428	PNN	RM 6 Loose log	1564	1708+vv	inc	MA limb	
DNT-429	PNN	RM 6 Loose log	1542	1697+vv	inc		
DNT-430	PNN	RM 5 Door lintel	1628	1711v	inc	MA shape	flat
DNT-431	PNN	RM 5 Roof primary	1545	1711L	inc	MA limb, bark, ends	SB
DNT-432	PNN	RM 5 Roof primary	1637	1712L	inc	MA limb, bark, ends	fork
DNT-433	PNN	RM 5 Roof primary	1602	1712LB	inc	MA limb, bark, ends	fork
DNT-434	PNN	RM 5 Roof primary	1571p	1711L	inc	MA limb, bark, ends	SB
DNT-436	PNN	RM 5 Loose log	1633p	1709+vv	inc	MA ends	
DNT-435	PNN	RM 4 Roof primary	1563	1710vv	inc	MA limb, bark, ends	
DNT-437	PNN	RM 4 Loose log	1589	1711vv	inc	MA bark,limb,shape,ends	flat

Key: DNT= Tree-ring Lab Collection; PNN= Pinyon; MA= Metal ax; limb= limb trim; bark= debarking; SB= Smoke blackening present; inc= incomplete

built at the same time or added later; the wall relationships between Rooms 1 and 2 and the main room-block have been destroyed and cannot be used to identify the temporal relationship between the two areas.

Tree-Ring Samples and Dates. Fifteen tree-ring samples, all of which produced dates, have been collected from the site (Table 14). Five of the samples, all from Room 5, produced cutting dates. All of the samples are pinyon, which is the dominant tree species in the immediate site area. All but one of the samples, a loose log in Room 6, show evidence of having been modified or procured using a metal ax. Two of the beams in Room 5, DNT-432 and 433, are forked poles.

Interpretations. Given the configuration of the main room-block (Rooms 4–7), the site must have been built during the pinyon growing season of A.D. 1712, probably during the spring or summer. Although a few of the beams in Room 6 may have been freshly cut (DNT-424, 425), some dead wood (DNT-426, 429) was also used in that room. Beams used in Room 5 include forked poles (DNT-432, 433) freshly cut in 1712, beams stockpiled from the previous year (DNT-430, 431, 434), and possibly a deadwood element (DNT-436). The dates from Room 4 (1710vv and 1711vv) come from eroded beams that could have been cut anytime after those dates, but were probably harvested during construction in 1712.

The Overlook Site (LA 10732)

The Overlook Site consists of a five-room masonry structure, four forked-pole hogans, and a masonry-based hogan near the edge of a cliff overlooking Palluche Canyon. The site is almost directly above Foothold Pueblito (LA 9073) and has an expansive view up Palluche Canyon (Figure 4). Another pueblito (LA 12929) may be visible, but Jacobson et al. (1992) do not indicate intervisibility between the sites.

Material Culture. Powers and Johnson (1987:63) indicate a ceramic assemblage dominated by Dinétah Gray and Gobernador Polychrome sherds, but Navajo Land Claims collections also include Zia, Tewa, Zuni-Acoma, and Hopi wares. No data concerning other artifact classes are available.

Architecture. The architecture indicates at least three distinct construction episodes at the pueblito (Towner 1997:243, Figure 61). Rooms 3 and 4 were apparently built as a single unit, Rooms 1 and 2 were abutted onto the existing structure as a unit, and Room 5 was added onto the west side of Room 2 at a later date. Room 4 contains two sealed doorways that may have been sealed when Rooms 1 and 2 were added. The existing doorway opens into Room 2, whereas the two sealed doorways opened into the Room 1 area. There is no architectural evidence to indicate the relative temporal position of the hogans.

Tree-Ring Samples and Dates. Thirty-two tree-ring samples, which yielded 27 dates, have been collected from the site (Table 15). The nine samples from the hogans, all collected by the NLC, yielded five dates, including one cutting, one near-cutting, and three noncutting dates. The 23 samples from the pueblito, collected during this project and by the NLC, produced 22 dates, including 16 cutting and 6 noncutting dates. All of the samples are pinyon, a somewhat unusual species distribution that may be partially a result of dendrochronological sampling bias. Of the 17 samples with true outsides and identifiable terminal ring characteristics, 13 were cut during the growing season and 4, all samples dating to 1727, were cut after the end of the growing season.

Table 15. Tree-ring Samples Collected from The Overlook Site (LA 10732)

Sample Number	Species	Provenience	Inside Date	Outside Date	Terminal Ring	Tool Marks	Comment
DNT-579	PNN	Room 4 Sealed Door lintel	1619	1727L	comp	MA limb, ends	
DNT-580	PNN	Room 4 Sealed Door lintel	1617	1727L	comp	MA limb, end	
DNT-581	PNN	Room 4 Sealed Door jamb	1651	1721v	inc	MA limb, burned end	added when door sealed (?)
DNT-582	PNN	Room 4 Roof primary	1666	1727LB	inc	MA limb	
DNT-583	PNN	Room 4 Roof primary	1512	1728L	inc	MA limb, ends	fork
NLC-1375	PNN	Room 4 ? Roof primary rubble	1609+p	1725	inc		
NLC-1376	PNN	Room 4 ? Roof primary rubble	1637+p	1727	comp		
NLC-1377	PNN	Room 4 ? Roof primary rubble	1571+p	1727G	comp		
NLC-1378	PNN	Room 4 ? Roof primary rubble	1554+p	1728	inc		
DNT-584	PNN	Room 3 Roof primary	1647	1727L	inc	MA limb, end	
DNT-585	PNN	Room 3 Door lintel	1646	1725L	inc	MA limb	
DNT-586	PNN	Room 3 Roof Auxilary	1659	1727LG	inc	MA ends	not cut to fit room
DNT-587	PNN	Room 3 Roof primary	1639	1727v	inc	MA bark, limb, end	
DNT-588	PNN	Room 3 Door Jamb	1655	1721vv	inc	MA end	
NLC-1372	PNN	Room 3 ? Roof primary rubble	NO DATE				
NLC-1373	PNN	Room 3 ? Roof primary rubble	1602+p	1727G	inc		
NLC-1374	PNN	Room 3 ? Roof primary rubble	1612+p	1725G	inc		
NLC-1366	PNN	Room 1 ? Roof primary	1603+p	1715+	inc		
NLC-1367	PNN	Room 2 ? Roof primary rubble	1634p	1715+			
NLC-1368	PNN	Room 2 ? Roof primary rubble	1624p	1726	inc		
NLC-1369	PNN	Room 2 ? Roof primary rubble	1645+p	1727G	inc		
NLC-1370	PNN	Room 2 ? Roof primary rubble	1600nc	1725	inc		
NLC-1371	PNN	Room 2 ? Roof primary rubble	1640p	1732	comp		
NLC-1379	PNN	Windbreak 3	1654+p	1742+G			

NLC-1380	PNN	Hogan 4 leaner	1641+p	1726	inc
NLC-1381	PNN	Hogan 4 leaner	1619+nc	1723	inc
NLC-1382	PNN	Hogan 6 leaner	NO DATE		
NLC-1383	PNN	Hogan 6 leaner	NO DATE		
NLC-1384	PNN	Hogan 7 leaner	1627+p	1682	inc
NLC-1385	PNN	Hogan 7 entry element	NO DATE		
NLC-1386	PNN	Hogan 7 fork	1611p	1727G	inc
NLC-1387	PNN	Hogan 7 fork	NO DATE		

Key: DNT= Tree-ring Lab Collection; NLC= Navajo Land Claim Collection; PNN= Pinyon; inc= incomplete; comp= complete; MA= Metal ax; limb= limb trim; bark= debarking

Interpretations. The Overlook Site provides several interesting interpretive possibilities. I believe that the most parsimonious explanation is as follows: Rooms 3 and 4 were built together during the 1727 growing season; evidence in the form of the incomplete terminal rings that exhibit no latewood cells on samples DNT-582, 584, 586, and 587 support this inference. Early in the spring of 1728, however, Room 4 was remodeled. The doorways were sealed and a new door opened. Evidence to support this inference is that the logs above the sealed doorway (DNT-579, 580) were cut after the 1727 growing season had ended, and the beam above the new door (DNT-583) was cut very early in the 1728 growing season. Another Room 4 beam (NLC-1378) was also cut in the spring of 1728 as evidenced by very few 1728 early wood cells on the specimen. I believe that Rooms 1 and 2 were added at the same time, primarily to protect the newly opened door. Room 5 was added later, but there is no dendrochronological evidence to indicate when that construction occurred. Sample NLC-1371 that dates 1731+vv may indicate Room 5 construction, but the provenience information indicates that it was taken from the rubble in Room 2. There is some evidence of stockpiling prior to the remodeling of Room 4 (DNT-585, NLC-1374, 1375) and also evidence of dead-wood use during the remodeling (DNT-581, NLC-1366, 1367). Hogans 4 and 7 were undoubtedly contemporaneous with the pueblito. The NLC sampled windbreak, however, was probably used after the pueblito was abandoned.

An alternative interpretation that has some dendrochronological support is that the entire structure was built in 1728 using freshly cut and stockpiled timbers. Architecturally, however, this seems unlikely.

Compressor Station (LA 5658)

Compressor Station is a four-room pueblito near a mesa edge high above the confluence of Largo and Palluche Canyons (Figure 4). Jacobson et al. (1992) indicate intervisibility with five other pueblito sites, but at least one of these (Foothold Pueblito) cannot be seen. The pueblito is part of a larger complex of sites, but is the only part of the complex that contains datable timbers. Cursory examination of the surrounding area by the author, Jeffrey S. Dean, and Christian E. Downum indicates that features associated with the pueblito include an isolated slab-lined hearth that was probably part of a hogan, a pile of burned rock indicative of a sweatlodge, a forked-pole hogan, and an alcove overhang that is approximately 25 centimeters deep with sheep dung and is the best preserved sheep corral associated with any of the pueblito sites. An alignment of small boulders along the front of the overhang demarcates the extent of the corral, but it contains no wooden elements useful for dating. As noted in Chapter 4, however, none of the dated corrals spatially associated with pueblitos are contemporary with the pueblitos. All are temporally associated with the Hispanic occupation of the Largo drainage in the late 1800s or early 1900s. Thus, the Compressor Station corral may indicate reuse by Hispanic herders long after the Navajo abandonment of the site.

Material Culture. Artifacts observed on the site include flaked stone and ground stone, faunal materials, and ceramics. The ceramic assemblage comprises Dinétah Gray and Gobernador Polychrome sherds. Single sherds of Jemez Black-on-white, Zia, Santa Ana, and Acoma types were noted during the Navajo Land Claim survey (Marshall 1991).

Architecture. There is no architectural evidence indicating the temporal relationship between the pueblito and the hogan or sweatlodge. Likewise, there is no way to know whether or not the sheep corral was used during the hogan or pueblito occupation, or at some other time. Wall analysis at the pueblito indicates that Room 1 was the first architectural unit constructed (Towner 1997:Figure 32), but it may not have been roofed. Rooms 2 and 3 were built next, probably as a single unit, and Room 4 was added after Rooms 2 and 3 were complete. Doorways between Rooms 1 and 2, 2 and 3, and 3 and 4 were sealed and new doorways opened some time after wall construction was complete.

Tree-Ring Samples and Dates. Thirteen samples, all of which yielded dates, were recovered from the site (Table 16); three of the samples (DNT-388, 389, 390) yielded cutting dates, seven yielded near-cutting dates (DNT-382, 383, 384, 386, 391, 393, 394), and three yielded noncutting dates (DNT-385, 387, 392). Pinyon was used for all of the roof beams and door lintels, but a juniper beam (not sampled) was used as a corner fireplace support. Most of the construction occurred during the growing season, a relatively rare situation at pueblito sites. All of the beams and lintels show evidence of having been cut with a metal ax.

Interpretations. There are at least three different interpretations of the data from the site. First, the entire pueblito may have been built in A.D.

Table 16. Tree-ring Samples Collected from Compressor Station (LA 5658)

Sample Number	Species	Provenience	Inside Date	Outside Date	Terminal Ring	Tool Marks	Comment
DNT-382	PNN	Roof primary RM 3	1610	1727v	comp	MA limb	
DNT-383	PNN	Roof primary RM 3	1660	1727v	comp	MA limb, ends	SB
DNT-384	PNN	Roof primary RM 3	1493	1728v	inc	MA limb, ends	SB
DNT-385	PNN	Roof primary RM 3	1613	1722++vv		MA limb, ends	spiral
DNT-386	PNN	Roof primary RM 3	1650	1727v	inc	MA limb, ends	SB; fork
DNT-387	PNN	Roof primary RM 3	1521	1724++vv	inc	MA limb, ends	SB
DNT-389	PNN	Door lintel RM 3/4	1603p	1727G	inc	MA limb, ends	
DNT-388	PNN	Door lintel RM 3/4 (sealed)	1654	1728G	inc	MA ends	
DNT-390	PNN	Roof primary RM 2	1595	1728G	inc	MA bark, limb, end	SB
DNT-391	PNN	Roof primary RM 2	1648	1727v	inc	MA limb	
DNT-392	PNN	Roof primary RM 2	1615	1727vv	inc	MA bark, limb, ends	
DNT-393	PNN	Door lintel RM 2/3 (sealed)	1577	1727v	inc	MA bark, limb, ends	
DNT-394	PNN	Door lintel RM 1/2 (sealed)	1621	1727v	inc	MA limb, ends	

Key: DNT= Tree-ring Lab Collection; PNN= Pinyon; inc= incomplete; comp= complete; MA= Metal ax; limb= limb trim; bark= debarking; SB= Smoke blackening present; spiral= spiral gran

1727 as a single construction episode, or a series of shortly spaced episodes. The cluster of 1727 cutting dates from the roofs of Rooms 2 and 3 and the door lintels support such an interpretation, although their relationship to Room 1 construction cannot be ascertained. If the structure was built in 1727, repairs in both rooms and sealing of the first doorway between Rooms 3 and 4 must have been done in the following year (1728). The repairs may have included the use of dead wood (DNT-385, 387) and freshly cut timbers (DNT-384) in Room 3 and Room 2 (DNT-390). They also must have involved the reuse of a previously cut door lintel (DNT-389) in the construction of the existing doorway between Rooms 3 and 4 and use of a freshly cut lintel (DNT-388) in the sealed doorway between Rooms 3 and 4.

The second interpretation is that the entire structure, with the possible exception of Room 1, was built in the spring of 1728. Cutting dates with incomplete terminal rings for that year are present in both the roofed rooms and the sealed door lintel between Rooms 3 and 4. If the structure was built in 1728, construction must have involved the use of freshly cut timbers (DNT-384, 388, 390), dead wood (DNT-385, 387), and logs (DNT-386, 389, 391, 392, 393, 394) stockpiled from the previous year. Beams DNT-382 and 383, although dated to 1727, were cut after the end of the growing season and thus may have been harvested prior to the 1728 growing season for use during construction.

The third possible interpretation is more complex and involves more assumptions concerning wood reuse. Architecturally, it is clear that Room 1 was built first; if Room 1 was roofed, none of the primary beams remain. If the original roof was removed from Room 1 when Rooms 2 and 3 were built, Room 1 may have been the source of roof timbers for the other two rooms. These logs were probably cut in 1727 for use in Room 1 and reused in Room 2 (DNT-391, 392) and Room 3 (DNT-386). If beams DNT-382 and 383 were actually cut in the spring of 1728 prior to the commencement of the growing season, there would be three primary beams in Room 3 that were cut in 1728 and only one that was cut in 1727. In addition, the sealed doorway between Rooms 3 and 4 also dates to the 1728 growing season. Because Room 2 contains a primary beam cut in 1728 and is architecturally bonded to Room 3, under this scenario it must have been built in 1728 also. Sealing of the old interior doorways and construction of new doors must have occurred during or after 1728 by reusing previously cut elements. Finally, Room 4 construction during or after 1728 must have been accomplished in conjunction with the sealing of the doorway between Rooms 3 and 4, but after Rooms 2 and 3 were constructed.

Each of these interpretations contains strengths and weaknesses. The first scenario, construction in 1727 and repair in 1728, relies on the strength of the 1727 date cluster and the distribution of those dates throughout the dated rooms. It provides no inferences for Room 1 construction, however, and cannot explain why such a structure should have been so extensively repaired after only one year of use. The second scenario, construction in 1728 using freshly cut beams, dead wood, and stockpiled beams explains the date distribution both temporally and spatially relatively well, but neither provides data relative to Room 1 construction nor explains why the doorway between Rooms 3 and 4 was built and sealed in the same year. The third interpretation, construction of Room 1 in 1727, removal of the Room 1 roof for construction of Rooms 2 and 3 with reused timbers, freshly cut beams, and dead wood in 1728, and construction of Room 4 later in the same year, explains the date distribution well, but suffers from a lack of evidence in Room 1. Of these three different possibilities, the second scenario seems the most probable. It is clear, however, that the site was built during the late 1720s and was probably occupied for only a short time.

Gould Pass Pueblito (LA 5659)

Gould Pass Pueblito is a two-room masonry structure and two associated forked-pole hogans located on a prominence between the Cereza and Largo drainages (Figure 4). It has excellent visibility, especially to the south towards Crow Canyon and northeast down Adolpho Canyon and into Adams Canyon. Jacobson et al. (1992) indicate that only the Cabresto Mesa Tower Complex is visible from Gould Pass Pueblito, but I suspect that Gomez Canyon, Gomez Point, and several other sites are also visible.

Material Culture. Powers and Johnson (1987) noted midden areas near the pueblito and hogans. Artifacts present consist mostly of Dinétah Gray ceramics, a few pieces of flaked stone, and rare pieces of bone. No Puebloan tradewares have been noted at the site.

Architecture. The masonry pueblito consists of two rooms on top of a small boulder (Towner 1997:205, Figure 47). A small jacal wall shields the exterior doorway from the remainder of Room 2; loopholes are present in the south and north walls and face toward Crow and Adolpho Canyons, respectively. Powers and Johnson (1987:45) indicate that the structure was built in two discrete episodes. A closer examination, however, shows that the walls of Rooms 1 and 2 are bonded and connected by a doorway; thus, the entire structure was built as a single construction episode.

Tree-Ring Samples and Dates. Twenty samples were collected from the site, 18 from the masonry structure and associated loose logs, and two from Hogan 2. None of the other hogan wood was suitable for sampling. Samples from all of the roof primary beams were collected, as were samples from door lintels, the jacal wall, and loose logs around the pueblito. Only three samples, all loose logs, did not date (Table 17). All of the visible beam ends and interior limb stubs have been cut with a metal ax. Six of the samples are pinyon and 14 are juniper, a species distribution similar to the local environment.

Interpretations. There is no way to determine the relationship between the hogans and the masonry structure because the 1710vv date from Hogan 2 is not a cutting date; the hogans may or may not be contempo-

Table 17. Tree-ring Samples Collected from Gould Pass Pueblito (LA 5659)

Sample Number	Species	Provenience	Inside Date	Outside Date	Terminal Ring	Tool Marks	Comment
DNT-852	JUN	RM 2 Roof primary	1568	1751GB	inc	MA flat, ends, limb	SB
DNT-126	JUN	RM 2 Roof primary	1554	1751LB	comp	MA ends, limb, bark	SB
DNT-127	JUN	RM 2 Roof primary	1563p	1751LGB	comp	MA ends, limb	SB
DNT-128	PNN	RM 2 Roof primary	1611	1751LGB	inc	MA ends, limb, bark	SB
DNT-129	PNN	RM 2 Roof primary	1621	1750+G	inc	MA ends, limb	SB
DNT-130	PNN	RM 2 Roof primary	1637	1749+LGB	inc	MA ends, bark	SB
DNT-131	PNN	RM 2 Roof primary	1430	1679++vv	inc	MA ends, bark	weathered
DNT-135	JUN	RM 2 Jacal Upright	1598	1753LGB	comp	MA limb	SB
DNT-136	JUN	RM 2 Jacal Upright	1641+/-	1749LGB	comp	MA end, limb	SB
DNT-137	JUN	RM 2 Jacal Upright	1607p	1749GB	comp	MA end, limb	SB
DNT-138	JUN	RM 2 Jacal Upright	1593+/-	1749GB	inc	MA limb	SB
DNT-139	PNN	RM 2 Door lintel	1456+/-	1706vv	inc	MA end	flat
DNT-132	JUN	RM 1/2 Door lintel	1615p	1750LB	inc	Indeterminate	SB
DNT-133	JUN	RM 1/2 Door lintel	1643	1747+L	comp	Indeterminate	
DNT-134	JUN	RM 1/2 Door lintel	1626p	1745vv	inc	MA ends	flat
DNT-141	JUN	Loose Hogan 2 fork	1461+/-	1710vv	inc	MA end, limb	
DNT-142	JUN	Loose Hogan 2 fork	NO DATE	—		MA end, limb	
DNT-140	PNN	Loose log (RM 1?)	1490	1746vv	inc	MA end, bark	flat
DNT-143	JUN	Loose log (RM 1?)	NO DATE	—		MA end	burned
DNT-144	JUN	Loose log Area 2	1488+/-	1692vv	inc	MA end, limb	
DNT-145	JUN	Loose log Area 1	1646	1749vv	inc	MA end	
DNT-146	JUN	Loose log Area 1	NO DATE	—		MA end	

Key: DNT= Tree-ring Lab Collection; JUN= Juniper; PNN= Pinyon; MA= Metal ax; limb= limb trim; bark= debarking; inc= incomplete; comp= complete; SB= smoke blackening present; flat= flattened

raneous. I assume, given the lack of contrasting evidence, that the hogans were occupied at the same time as the masonry structure.

There are at least two valid interpretations of the site construction history. First, the entire structure *may* have been built (but not roofed) in the late summer or early fall of 1749 when the jacal wall was erected. Indeed the masonry may have been built at any time, but it is extremely unlikely that the jacal wall was built before the structure was in place. Next, the structure was roofed in the late summer/early fall of 1751 using freshly cut beams (DNT-852, 126, 127, 128 and probably 129), a stockpiled or reused beam (DNT-130), and a piece of dead wood (DNT-131). Finally, the jacal wall was repaired in the fall of 1753 or spring of 1754 before the growing season started.

The second, and more likely, scenario is that the entire structure was built in the late summer/early fall of 1751. Some beams were cut specifically for the construction, and some dead wood was used in the roof, as door lintels, and as part of the jacal wall. The jacal beams *may* have come from one of the hogans, but there is no specific evidence to support such an inference. The jacal structure was repaired after the 1753 growing season ended, and the roof may have been repaired at the same time using dead wood (DNT-131); the only evidence for roof repair is the lack of smoke blackening on sample DNT-131, but it is far from conclusive.

Finally, it is possible that the pueblito was built in 1749 and reroofed in 1751 using freshly cut logs and elements from the original roof. Both the roof and jacal wall may have been repaired in 1753. This hypothesis suggests that the site was occupied episodically in 1749, 1751, and 1753, with periods of abandonment between construction or repair events. In any case, the site was probably occupied only very briefly, probably between A.D. 1751 and 1754 at the latest. The loose logs contribute little to the site chronology, except to reinforce the inference of no occupation past the 1753 growing season.

Hooded Fireplace Ruin (LA 5662)

The Hooded Fireplace Ruin is located south of Largo School Pueblito (LA 5657) on a wide bench above Largo Canyon (Figure 4). Jacobson et al. (1992) indicate that the Cabresto Mesa Tower Complex (LA 2138) and Casa Mesa Diablo (LA 11100) are visible from Hooded Fireplace Ruin. Marshall (1991:33–51) identified six middens, three masonry-based hogans, a possible corral, and a possible ramada adjacent to the pueblitos. The larger site complex included additional middens, hearths, and deteriorated boulder-backed masonry rooms. The pueblito consists of six masonry rooms: four contain intact roofs or roof segments, one (Room 3) has an intact Spanish-style "hooded" fireplace, and another (Room 2) contains the remains of a similar fireplace. The site is high on the bench and overlooks a wide expanse of the Largo drainage.

Material Culture. Powers and Johnson (1987:52) noted Dinétah Gray and Gobernador Polychrome ceramics, flaked and ground stone artifacts, a wooden awl, and faunal remains. Marshall (1991) identified Puname Polychrome, Tewa Polychrome, and Ako Polychrome ceramics at the site, but such intrusive wares comprised less than ten percent of the identified ceramics during the survey and less than three percent of the ceramic assemblage documented during the data recovery phase of the project (Marshall 1995). Lithic materials included Jemez obsidian, but were dominated by cherts, quartzites, and chalcedony (Marshall 1995).

Architecture. The preliminary map of Powers and Johnson (1987) indicates that the entire pueblito was built as a single construction episode. They also suggest that three of the rooms may have had second stories, but the upper parts may have simply been "parapets." All of the visible wall elements in the structure are bonded, indicating a single construction event.

Tree-Ring Samples and Dates. A total of ten samples, nine of which yielded dates, was collected from this site. The dates (Appendix) are relatively tightly clustered in the early 1720s. The one sample that did not date (DNT-52) is from a cottonwood (*Populus spp.*) roof beam in Room 1. All of the datable beams and nine of ten samples taken from the site are pinyon. The selection of a single species for construction purposes is not seen in any other pueblito and is not a result of dendrochronological bias. Pinyon apparently met the size, length, and straightness criteria for construction of these square, bonded rooms.

Interpretations. The cluster of cutting dates from Room 3 indicate that this room was constructed in A.D. 1723. The additional A.D. 1723 cutting date from Room 1 and the cluster of noncutting dates from Room 2 indicate that these rooms were also built in A.D. 1723. It is possible that the noncutting A.D. 1723vv date from Room 4 indicates a later construction or repair event, but this seems unlikely. All but one of the samples with true outsides (bark, L-condition, or galleries) have complete terminal rings; the only exception is the shelf pole in Room 1 that was cut during the pinyon growing season. These data indicate that Hooded Fireplace Ruin was constructed in its entirety after the end of the A.D. 1723 pinyon growing season, probably in the late summer or early fall.

Based on the volume of midden deposits, Marshall (1995:211) estimates the site was occupied until A.D. 1763. The lack of tree-ring dates after A.D. 1723, combined with the architectural evidence of a single construction episode without additional repairs, suggest that the structure was occupied for only a short time. None of the pueblitos in the Dinétah was continuously occupied for 40 years; indeed, the latest tree-ring date of any type in the Dinétah, a firewood sample from a roasting pit, dates to 1762 (Fetterman 1996). In addition, Old Fort Ruin, one of the largest site complexes, was occupied for less than 10 years and Three Corn Ruin, another large complex, was used, abandoned, and reoccupied at least once during a span of approximately 20 years (Towner and Johnson 1998). Marshall's (1995) estimate, therefore, clearly overestimates the length of the occupation, probably by decades. I seriously doubt the site was occupied later than A.D. 1730, if it was occupied that long.

CHAPTER 4

Temporal and Spatial Patterns of the Pueblito Site Occupations

The tree-ring data provide an unparalleled opportunity to examine the Navajo occupation of the Dinétah from a number of perspectives. The key to conducting this type of analysis is to use the appropriate parts of the database to address different questions. In this chapter, I use the tree-ring data to date five specific target events (Dean 1978) and additional data to address the pueblito phenomenon geographically.

THE PUEBLITO OCCUPATIONS

Because of the large size of the tree-ring database, I am able to assess the pueblito occupation from a number of different perspectives, including tree-harvesting activities, building construction, and probable time of occupation. Once these analyses are complete, I then discuss the temporal distribution of pueblito occupation as it relates to various historical factors.

Tree-Ring Dates

The overall date distribution (Figure 26), including all 798 dates from pueblito sites that relate to the Navajo occupation, shows a relatively continuous distribution throughout the early to mid-1700s with large modes in the 1710s, 1720–1735, and 1745–49 periods. Because this distribution includes noncutting dates, however, it is undoubtedly skewed by those processes, both natural and cultural, that remove the outer rings from samples. Nevertheless, this distribution is informative, particularly

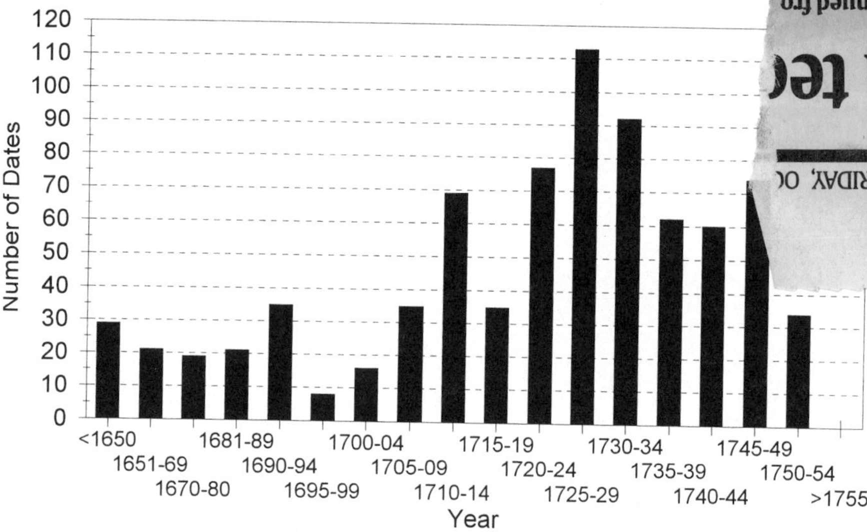

Figure 26. Temporal distribution of all pueblito site tree-ring dates (n = 798)

regarding its late end. The relatively abrupt decline in dates in the 1750s, particularly the lack of dates after 1755, suggests a sudden depopulation of the area, or at least a cessation of tree harvesting and pueblito construction.

Cutting and Near-cutting Dates

The cutting and near-cutting dates (n=311) from the sites (Figure 27) show quite a different distribution. These dates range from 1634L to 1754+G and include two significant date clusters prior to 1700, one at the Buffalo Mask site and the other at Tapacito Ruin. The pre-1670 cutting dates (n=5) are all anomalous in their site contexts. The 1634L date is from Room 2/3 at Three Corn Ruin—a room-suite that was built in 1714; the 1638G date is from Simon Canyon, which was built in 1754; the 1642v date is from The Citadel—a site that was built sometime after 1689; the 1652+LGB date is from the Lone Warrior site—a site built mostly with dead wood after 1682; and the 1655G date is from Rabbit House—a site built almost a century later in 1752. Although these particular specimens do not exhibit the missing rings characteristic of deadwood use, I believe all were procured after tree death. Therefore, none of these samples can be used to suggest pre-1670 tree-harvesting activities.

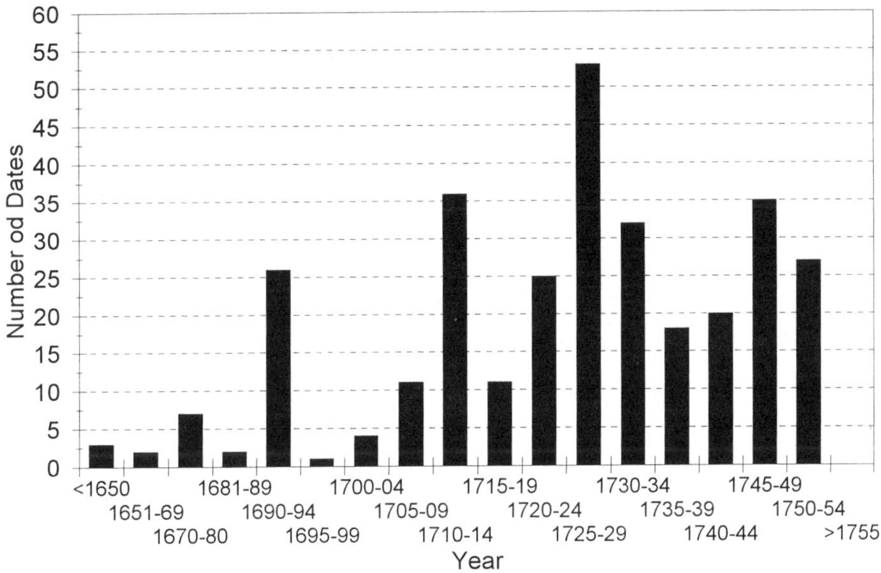

Figure 27. Temporal distribution of all pueblito site cutting and near-cutting dates (n = 311)

The 1670–1680 cutting and near-cutting dates (n=8) are derived from the Buffalo Mask site (n=6), Frances Canyon Ruin (n=1), and Morris Cliff Dwelling (n=1). The latter two dates are, like the pre-1670 dates, probably the result of dead-wood use. The 1672B date from Frances Canyon Ruin is from a beam in Room 4, constructed in 1722; the 1675B date from Morris Cliff Dwelling is from Room 8, which also yielded a 1720 cutting date. Thus, the only reliable evidence at the pueblito sites for pre-1680 tree procurement comes from the Buffalo Mask site, where six trees were harvested between 1670 and 1680.

The large cluster of cutting dates in the 1690–1694 period is the result of rapid construction, good preservation, and extensive sampling at Tapacito Ruin (Towner and Dean 1992). Of the 27 cutting or near-cutting dates in the period, 16 are from the pueblito at Tapacito Ruin and 9 are from an associated hogan; a single date of 1692G is from the ladder at the Buffalo Mask site, and a 1691G date is from Simon Canyon. Obviously, the large number of dates in the 1690–1694 period cannot be used to infer activities except at Tapacito and, to a minor extent, at Buffalo Mask.

The next large peak in the distribution occurs in the 1710–1714 period. The relative paucity of dates in the 1695–1709 period, however, may be important. Only a single cutting date from the 1695–1699 period exists; the 1699G date is from a hogan leaner at 42 Pueblito and its provenience is somewhat questionable.

There are four cutting or near-cutting dates in the 1700–1704 period. Two of the dates, 1701G and 1701+LB, are from the upper room-suite at Three Corn Ruin, which was built in 1714; one of the dates, 1701L, is from the door lintel between Rooms 1 and 3 at Garcia Canyon Pueblito, although the rooms were not built until 1712. The final date, 1701+LB, is from the only in situ beam at Tso'bito Pueblito and could date that structure; a loose log near the structure, however, dates 1726v and may also date the initiation of construction.

Eleven specimens date to the 1705–1709 period. Six of these dates are from the upper room-suite at Three Corn Ruin which was built in 1714; the large number of pre-1714 cutting dates from this provenience suggests that many of these beams were reused from earlier structures. Another cutting date, 1707c, was derived from Hogan 4 at Three Corn, but the same structure also yielded a 1715 cutting date. A hogan fork at Pointed Butte yielded a 1709c date, but the same structure had leaners that dated 1721 and 1748, so the fork may have been a reused or dead-wood element. Two near-cutting dates (1707+B and 1708++v) from McKean Pueblito indicate construction at that site. The only other date from the period, 1705GB, is derived from a loose log at Shándin Bíikin; it is the latest date from the site and the only cutting date, so, although it may date the structure, it is not very conclusive evidence of construction activities.

Following A.D. 1710, tree-harvesting activities increased dramatically. Thirty-six beams, representing activities at eleven different sites, were cut in the 1710–1714 period. At least six of these sites exhibit large clusters of cutting dates, including Three Corn, Frances Canyon, and Garcia Canyon, Kin Wólá Jíní, Rincon Rockshelter, and the Cabresto Mesa Tower Complex. The other four sites, McKean Pueblito, Shaft House, Delgadito Pueblito, LA 2297, and Pueblito East, have small numbers of dates and only one or two dates in the period. Delgadito Pueblito and LA 2297 also include later dates, so the dates from this time period may actually represent reuse or dead-wood harvesting. Reasons for the dramatic increase in cutting activity during the 1710–1714 period are discussed below.

There is a marked decrease in the number of trees cut in the 1715–1719 period (n=11). These samples are derived from seven sites. The occupations at three of the sites, Three Corn Ruin, Frances Canyon Ruin, and Garcia Canyon Pueblito, were apparently continuations from the previous period; one other site, LA 2297, has cutting dates in both periods, but no strong date clusters. Three new sites, LA 2305, Crow Canyon Pueblito, and Dooyiit'íínî Pueblito all contain cutting or near-cutting dates in the period, but no strong clusters, and all but Dooyiit'íínî Pueblito have later cutting or noncutting dates as well.

The resurgence in tree harvesting in the 1720–1724 period resulted in 25 cutting or near-cutting dates from nine sites. Tree-cutting activities continued at Three Corn, Garcia Canyon, and Frances Canyon Ruins, and commenced at Hooded Fireplace, Tower of the Standing God, Morris Cliff Dwelling, and Hummingbird House. The strong cluster of noncutting dates from Romine Canyon Pueblito suggests that it was built during this time period. A hogan at Pointed Butte may have been built, but the provenience information on the samples is poor and possibly in error. A single cutting date at Boulder Fortress is clearly anomalous and may represent either dead-wood or beam reuse, probably the former.

The largest number of trees cut during the pueblito occupation of the area was procured during the 1725–1729 period. Fifty-three cutting or near-cutting dates derived from ten sites comprise the largest mode in the distribution. Tree-harvesting activities continued at Three Corn Ruin, Boulder Fortress, and Hummingbird House, but the remaining sites appear in the database for the first time. Large-scale tree harvesting occurred at Canyon View Ruin, Compressor Station, The Overlook Site, Kin Naa dáá', and possibly Dos Cerritos. In addition, the strong cluster of noncutting dates from Twine House indicates that it was built during the late 1720s. Tree procurement also may have taken place at Santo Niño (1726r), but the single date may be the result of dead-wood use; the single date at Tso'bito Pueblito (1726v), however, is the only date from the site and may actually date construction or repair activities.

The 1730–1734 period saw a decline in tree procurement (n=32), although only a slight decline in the number of sites with evidence of tree cutting. Nine sites indicate some activity; tree cutting continued at Three Corn Ruin, Canyon View Ruin, and Dos Cerritos. Jaramillo Canyon, Hill Road, Gomez Canyon, and Foothold Pueblitos show evidence of tree cutting, as do Truby's Tower and the Pork Chop Pass site. Of these sites that indicate new activity, however, only Jaramillo Canyon was constructed during this interval.

The 1735–1739 period shows a continued decline in the number of trees cut (n=18), but the number of sites with evidence of cutting activity is similar (n=10). Tree harvesting continued at Three Corn, Hill Road, and Garcia Canyon, as well as at Pork Chop Pass; it commenced at Old

Fort, Kin Yazhi, Largo School, Quarter-Section Pueblito, and Pueblito Canyon Pueblito. Tree cutting also resumed at Frances Canyon Ruin during this period. Most of these sites show substantial date clusters, although those at Pueblito Canyon, Old Fort, Largo School, and Pork Chop Pass are small.

Only twenty trees were cut in the early 1740s (1740–1744), but represent activities at ten sites. Six sites (Old Fort, Frances Canyon, Hill Road Ruin, Kin Yazhi, Quarter-Section, and Pork Chop Pass) suggest that activity continued from the previous period. Two sites contain clearly anomalous dates (Hummingbird House and Truby's Tower), one date represents reuse (The Overlook Site), and two sites (Hidden House, Munoz Canyon Pueblito) appear in the database for the first time.

The late 1740s (1745–1749) saw an increase in the number of trees cut (n=35) and an increase in the number of sites (n=13) with evidence of tree harvesting. Six sites show that tree harvesting continued from the previous period, and four sites (Morris 1, Adolpho Canyon, Gould Pass Pueblito, and Santo Niño) appear in the database for the first time.

The last time period in the database, 1750–1754, exhibits evidence of tree harvesting at seven sites represented by 27 cutting or near-cutting dates. Tree cutting continued at Old Fort, Gould Pass, Santo Niño, Hidden House, and Truby's Tower and was initiated at Simon Canyon and Rabbit House. The distribution of cutting and near-cutting dates from pueblito sites, indeed all dates from these sites, ends abruptly in 1754. There are no later dates from any of the pueblitos in the Dinétah, although later dates from pueblitos outside the area do occur (Brugge 1972; Towner 1996).

Summary of Tree-Cutting Activities

The cutting dates provide evidence that the Navajo have occupied areas around the pueblito sites since the mid- to late-1600s. Although there are a few pueblito dates in the 1630–1670 period, most are anomalous in their archaeological contexts and are probably dead-wood elements or elements reused from earlier hogans. Beginning in the 1670s, however, clusters of cutting dates indicate more intensive use of the area. Following the tree-cutting activities at the Buffalo Mask site, a major procurement event occurred at Tapacito Ruin in the early 1690s. Almost all samples that date to this period are from Tapacito, and they constitute a major mode in the cutting date distribution.

The paucity of cutting dates in the 1695–1709 period suggests little activity. It is important to note that during this interval, the purported time of a "refugee" influx, a total of 14 trees was harvested from the areas around seven sites. The majority of these trees (n=9) were used at Three Corn Ruin, a possible indicator that the area around it was previously occupied. The small number of cutting dates in the 1695–1709 period hardly seems to be an indicator of a "massive" influx of immigrants. Three of the fourteen samples are from hogans, but may not date the initiation of hogan construction. Single dates from hogans, even cutting or near-cutting dates, are generally poor indicators of construction because of the Navajo tendency to reuse beams (Oswald 1993). The only positively identified construction during this period occurred at McKean Pueblito. Notably, no other sites in northwestern New Mexico, neither pueblitos nor forked-pole hogans, can be dated precisely to this interval. Thus, the concept of a massive influx of refugees immigrating to the Dinétah following the failed revolt of 1696 has virtually no support in the archaeological record.

Tree cutting seems to have occurred in spurts during the occupation of the area. Relatively large numbers of trees were procured in the early 1710s, from 1720 to 1734, and in the late 1740s and early 1750s. Tree cutting virtually ceased near the pueblitos after 1754, suggesting that the area was depopulated rather suddenly in the middle of the 1750s.

Pueblito Construction Episodes

Another way of examining the tree-ring data is to view the dates, including cutting, near-cutting, and noncutting dates, in their archaeological contexts. As described in Chapter 3, using this method I defined "construction episodes" at individual sites. The method used here differs slightly from that employed in Towner (1996); in that publication, I used cutting-date clusters, strengthened by noncutting dates, without a detailed architectural analysis. For example, I defined a "construction episode" in 1709 at Three Corn Ruin based on the two 1709 cutting dates in Room 2/3 (Towner 1996:159). As indicated in Chapter 3, however, Room 2/3 could not have been built prior to 1714. When the construction episodes defined on tree-ring and architectural criteria are viewed in the aggregate (Figure 28), a distinctive picture of pueblito construction emerges.

The Buffalo Mask site, constructed in A.D. 1680, is clearly the earliest structure of its type and remains something of an enigma. Its location high

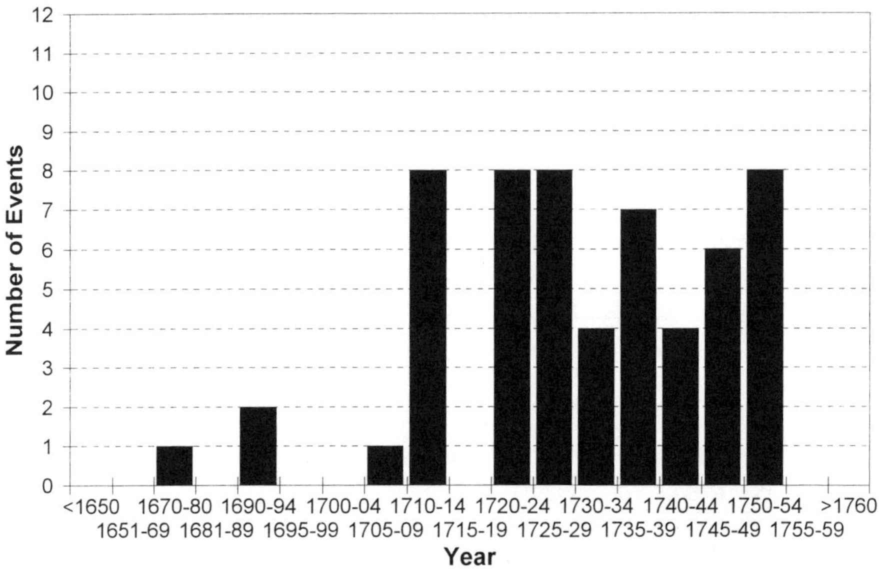

Figure 28. Temporal distribution of pueblito site construction episodes (n = 57)

in an alcove is clearly defensive, and the scarcity of material remains suggest it was indeed a defensive refuge, not a habitation. Although early Navajo pictographic "ceremonial imagery" (Copeland and Rogers 1996) is located nearby, there are no hogans or other evidence of habitation in the immediate area, and the site and rock art are not intervisible. Copeland and Rogers (1996:222) suggest that the imagery predates the Pueblo Revolt; if so, it may be contemporaneous with the rockshelter site. Buffalo Mask was built at least 30 years earlier than any other rockshelter site, such as Rincon Rockshelter or Shaft House. Brugge (personal communication, 1994) suggests that it represents construction by an early group of refugees who escaped the Rio Grande *during* the Pueblo Revolt of August 1680. Because all the terminal rings on the 1680 and 1679+G samples are incomplete, if the site was built by refugees, they must have left the Rio Grande almost immediately to reach the Dinétah before the end of the pinyon and juniper growing seasons in August or September. I do not agree with this hypothesis, however. The Puebloans were victorious in 1680 and had no need to flee; those Puebloans who sided with the Spaniards most likely would have fled southward toward Isleta and El Paso, not north into the Navajo country. Nevertheless, the Buffalo Mask site remains problematic in terms of its relationship to other pueblito sites in the Dinétah.

The next construction episodes occurred at Tapacito Ruin near the confluence of Tapacito Creek and the Largo. Towner and Dean (1992) argue that the 1690 cutting dates were derived from a hogan associated with the masonry structure, which was built in 1694. With massive core-veneer walls, no external doorways, a roof entryway, a distinct ceramic assemblage (Marshall 1991, 1995), and its situation on a structural bench rather than on a boulder or mesa top, Tapacito Ruin is unique among the pueblitos. Brugge (1996:262) believes that its lack of defensive orientation is one of the most important aspects of the site. Although he suggests that it has lines-of-sight to several other sites, Jacobson et al. (1992:130–31) indicate that only Split Rock and Truby's Tower are visible. My field checks indicate that Split Rock is not visible, however, and even so, probably was not built until the mid-1720s. Truby's Tower was not built until at least 50 years later. Thus, Tapacito was an isolated site during the entirety of its occupation. On a recent visit, two traditional Navajos asserted that the structure is not Navajo, primarily because it has no east-facing exterior doorway. Tapacito Ruin "may be the only genuine Pueblo refugee structure in the Dinétah" (Towner and Dean 1992:327).

Towner, Rogers, and Copeland (2001) suggest that the building "boom" of the early 1710s was a response to Roque de Madrid's campaigns against the Navajo. Construction of living spaces occurred at Garcia Canyon Pueblito (1712), Kin Wólá Jíní (1712), Cabresto Mesa Tower (1713), McKean Pueblito (1713), and the upper room-suite at Three Corn Ruin (1714), and storage rooms were built at Frances Canyon Ruin (1710, 1714) and probably Rincon Rockshelter (1713). In addition to these structures, two other pueblitos may have been occupied in the early 1700s; the dates at Three Corn suggest an earlier (1705–1709) structure, and historical documents (Hendricks and Wilson 1996:81) indicate Los Peñoles (LA 8948) was used as a refuge in 1705.

Although material remains at Los Peñoles are scarce, it may be important for other reasons. When the Spaniards under Roque de Madrid's command attacked the site on August 12, 1705, the Navajos defended the height and "won a tactical victory" (Hendricks and Wilson 1996:81). Although I believe most pueblitos were built for protection against Ute raiders (Towner 1996:166), the successful defense of Los Peñoles against the heavily armed Spaniards may have provided a "prototype" settlement for other pueblitos. This may help explain the construction boom of the early 1710s, a time when Ute raiding was not yet common, but when the Spaniards were conducting campaigns against the Navajo, in 1709 (n=5), 1713, 1714, and 1716 (Hendricks and Wilson 1996:98–100).

The absence of pueblito construction in the late 1710s suggests that the threats to the Navajo had subsided. Clearly, the Spanish threat had abated, but historical documents (Hill 1940) suggest that Ute raiding increased at the same time. Although some trees were harvested in the 1715–1719 period (see above), no construction episodes can be positively identified.

The early 1720s saw a renewed interest in pueblito construction with eight building events at seven sites. Interestingly, only at Frances Canyon and Garcia Canyon was construction part of a renewed activity; the other six sites were apparently founded during this interval. I have argued elsewhere (Towner and Johnson 1998) that Three Corn Ruin was abandoned during the early 1720s, and it is possible that the Three Corn inhabitants moved across Gobernador Canyon to Romine Canyon Pueblito.

Newly constructed sites were also the norm in the late 1720s. At least eight construction episodes can be defined, all but one of which are at sites without previous construction. The exception is the construction of the lower room-suite at Three Corn Ruin. Thus, during the 1720s construction was initiated at twelve new sites with only Three Corn, Frances Canyon, Garcia Canyon Pueblito and The Overlook Site showing any evidence of repair or remodeling.

The construction decline in the early 1730s is made more apparent by the fact that one of the four building events was a remodeling of the lower room-suite at Three Corn. Only at three sites, Jarmillo Canyon, Dos Cerritos, and Canyon View Pueblito, was construction initiated during the early 1730s. The late 1730s indicate a different strategy. Of the seven identified construction events, three are at sites that may have been previously abandoned. Construction at Three Corn continued with the reoccupation of the upper room-suite, new rooms were added at Frances Canyon Ruin after a hiatus of more than 15 years, and additional rooms were added to Deer House more than 10 years after the site was founded. Construction was also initiated at Gomez Canyon, Largo School, Quarter-Section Pueblito, and Hill Road Ruins between 1735 and 1739.

Only one site, Muñoz Canyon Ruin, was founded in the 1740–1744 period, and only four construction episodes can be identified; although Twine House, 42 Pueblito, and The Overlook Site may have been reoccupied (Ababneh et al. 2000). The three other building events were remodeling events at Frances Canyon, Quarter-Section, and Hill Road Ruins. In contrast, all four of the building events in the late 1740s occurred at newly founded sites. Two major building events occurred at

Old Fort Ruin, and both Hidden House and Morris Site 1 were founded.

Pueblito construction increased dramatically in the early 1750s. Three major building events occurred at Old Fort, and another occurred at Hidden House. But the four other events occurred at sites with no evidence of previous construction—Gould Pass Pueblito, Rabbit House, Truby's Tower, and Simon Canyon. This construction is something of a paradox because all of the sites were abandoned shortly thereafter, almost certainly by the late 1750s. There are no additional pueblito construction episodes after Simon Canyon was built in 1754.

Hogans at Pueblito Sites

One of the important questions concerning the pueblito sites has been the temporal relationship between the masonry pueblitos and forked-pole hogans. Kidder (1920) implicitly suggested that the hogans and pueblitos were simultaneously occupied, the hogans by Navajos and the pueblitos by Puebloan refugees. There have been some suggestions (Marshall 1995:204; Towner and Dean 1992) that hogans were occupied earlier than pueblitos and represent either (a) different populations, or (b) different settlement patterns not yet affected by defensive requirements (Brugge, in press).

Although the data are somewhat limited, there are several ways to evaluate the contemporaneity of the two structure types. The distributions of hogan and pueblito tree-ring dates are very similar, although the number of hogan dates is only about one-tenth that of the pueblito dates. The distributions of hogan and pueblito cutting dates (Figure 29) are also very similar; however, only 21 cutting or near-cutting dates were derived from hogans. A comparison of pueblito and hogan construction episodes would be an ideal way to determine structure contemporaneity. Unfortunately, only Tapacito Ruin and Hidden House contain pueblitos and hogans with identified construction episodes. The putative hogan at Tapacito was built four years prior to the pueblito construction, and the hogan at Hidden House was built as the masonry structure was expanded and remodeled. Marshall's (1996:209) ceramic analysis suggests that "the hogans at both Tapacito and Hooded Fireplace Pueblitos may slightly pre-date the pueblitos."

The most productive way to assess structure contemporaneity is by examining the sites with both hogans and masonry structures. There are 21 dated pueblito sites that contain hogans (Table 18 and Appendix); two of

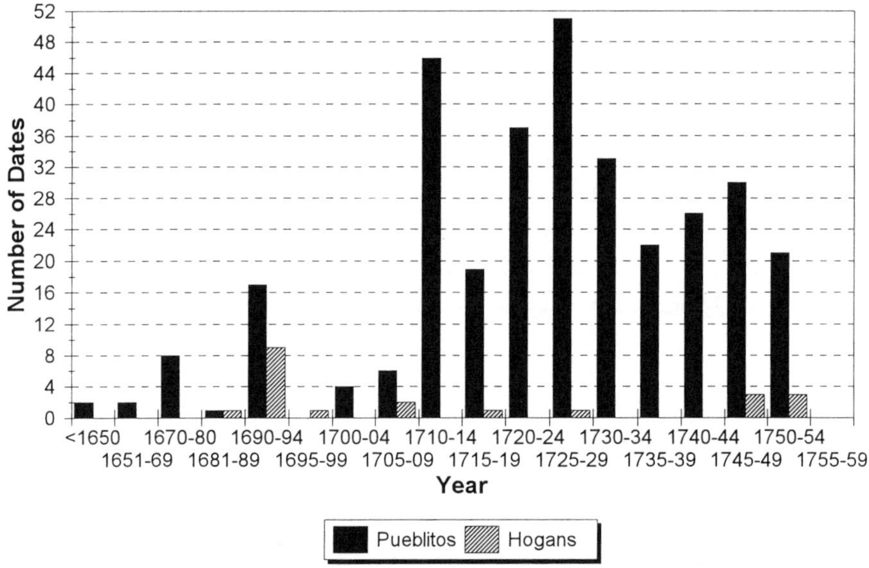

Figure 29. A comparison of hogan and pueblito cutting and near-cutting dates

the sites, Cottonwood Divide and Unreachable Rockshelter, contain only hogan remains, and one site, Dooyiit'íiní Pueblito, contains a brush structure that may or may not be part of a hogan. Tree-ring samples have been collected from both structure types at the remaining sites. Although at twelve of the sites the hogan dates are earlier than the pueblito dates, ten of the sites contain only noncutting dates from hogans; only two sites, Tapacito Ruin and 42 Pueblito, have hogan cutting dates earlier than the pueblito cutting dates. Unfortunately, no construction episodes have been positively identified at 42 Pueblito and the single hogan date may be anomalous. Thus, only at Tapacito Ruin can I demonstrate that a hogan existed prior to pueblito construction.

At seven of the sites, including four of the six sites with cutting dates from hogans, the hogans appear to be either contemporary with or later than the masonry structures. This assessment is based on the evidence that the hogan dates are either in the same year as pueblito construction or cutting dates, or are later than the latest pueblito date. In addition, at seven of the hogan sites with only noncutting dates, those dates are less than 50 years prior to pueblito construction. Given the amount of ring loss indicated by the "vv" symbol on most dates, I suspect that these seven hogans are contemporary with pueblito construction. There are only two

Table 18. Comparison of Forked-pole Hogan and Pueblito Dates at Pueblito Sites

Site Name	Pueblito Construction	Latest Pueblito Date	Number of Dated Hogans	Number of Hogan Dates	Hogan Date Range	Comment on Hogan Dates
Adolpho Canyon Ruin	Unknown	1747	3	5	1725–1751	Latest Date Cutting
Adams Canyon Ruin	>1736	1736	1	1	1630	Noncutting Date
Three Corn Ruin	>1714	1745	1	3	1668–1715	Latest Date Cutting
Old Fort Ruin	>1746	1753	7	12	1651–1748	All Noncutting Dates
Frances Canyon Ruin	>1710	1745	1	3	1566–1589	All Noncutting Dates
42 Pueblito	1727–1741?	1741	1	1	1699	Cutting Date
Boulder Fortress	1728	1728	1	1	1709	Noncutting Date
Canyon View	1734	1734	1?	3	1601–1725	All Noncutting Dates
Cottonwood Divide	N/A	N/A	4	7	1624–1715	All Noncutting Dates
Dooyiit'ííní Pueblito	>1719	1719	1	1	1712	Noncutting Date
Foothold Pueblito	>1739	1740	1	4	1667–1710	All Noncutting Dates
Gomez Canyon Pueblito	>1735	1747	4	7	1681–1733	All Noncutting Dates
Gould Pass Pueblito	>1749	1753	1	1	1710	Noncutting Date
Hanging Pot	>1747	1747	1	1	1735	Noncutting Date
Hidden House	>1746	1754	1	6	1735–1751	Late Cutting Date Cluster
Morris Cliff Dwelling	1720	1720	1	2	1726	Noncutting Date
The Overlook Site	>1727	1728	2	4	1682–1727	Latest Cutting
Pointed Butte	>1713	1713	3	6	1683–1748	Latest Date Noncutting
Split Rock Ruin	>1727	1727	2	2	1570–1665	All Noncutting Dates
Tapacito Ruin	1694	1694	1	8	1690	Cutting Date Cluster
Unreachable Rockshelter	N/A	N/A	4	6	1633–1680	All Noncutting Dates

sites with noncutting hogan dates, Adams Canyon and Split Rock Ruin, that are more than 50 years older than the pueblito dates. Although at some future date it may be possible to estimate cutting dates for these samples (cf. Nash 1997), such an approach has not been attempted here. Nevertheless, given the absence of sapwood on most hogan samples that yield noncutting dates, I believe the evidence indicates that the hogans and masonry structures on these sites were constructed at the same time and represent a single population using both structure types.

Pueblito Site Occupation Spans

Estimating the occupation span of the pueblito sites is critical for delineating the population dynamics of the Dinétah during the eighteenth century. One of the implicit assumptions of archaeologists and historians working in the area has been that these sites were occupied for most, if not all, of the Gobernador Phase (A.D. 1690–1775). Hester (1962a:86) used the number of hogans and pueblitos occupied in the Navajo Reservoir District, Largo and Gobernador "localities," and at Big Bead Mesa

(Keur 1941) to estimate a Navajo population of approximately 2,000 individuals (including Puebloan refugees) in A.D. 1700. Because his figures generally agreed with the Rabal Document (Hill 1940), they have been used as a baseline for reconstructing Navajo population dynamics.

Assuming that all pueblitos were occupied at the same time, however, is clearly unwarranted. The temporal distribution of tree-ring cutting dates and pueblito construction episodes demonstrates that many structures were built in the last few decades of the Dinétah occupation. It may be more important, however, to realize that once built, pueblitos may have been abandoned, reoccupied, and abandoned again within relatively brief periods of time. If such settlement shifts were the norm, a very different picture of eighteenth-century Navajo population dynamics must be constructed.

Several different methods are available for estimating site occupation spans. Robinson and Towner (1993) provide some information relevant to estimating occupation duration by listing the earliest cutting and latest date from each site and structure type. Although somewhat useful, these data cannot account for beam reuse or recycling, or dead-wood samples that yielded "cutting" dates. Hantman (1983:111–23, 143) advocates subtracting the date of initial construction from the latest tree-ring date and adding a constant of 10 years to determine the occupation span for pit house and pueblo sites. Although this and other methods are tested below, it is apparent that such an approach is inappropriate for Navajo structures because they are short-lived structures. Alhstrom (1985: 658–59) suggests that such an approach may be warranted if there is a continuous date distribution *and* if the latest occupied portions of a site yield tree-ring dates. The latter criterion is usually not a problem at the pueblito sites because the sites are generally small and all samples appropriate for dating have been collected. The first criterion, however, is generally met only at the larger sites with more than 10 or 20 dates.

Marshall (1995:209–12) uses a midden-accumulation-per-room-unit rate based on a 45-year occupation of Frances Canyon Ruin. Although he acknowledges several problems with his formula, Marshall uses ".45 cubic meters of trash fill per room per year" to estimate the occupation spans of seven pueblitos. His formula resulted in estimates of 3- and 4-year occupations at Shaft House and Split Rock Ruin; 10- and 11-year occupations at Largo School and Crow Canyon; a 20-year occupation of Tapacito Ruin; and 40- and 45-year occupation spans at Hooded Fireplace and Frances Canyon Ruins respectively.

Marshall's approach has been criticized by Towner and Johnson (1998:176–79) for ignoring the cultural and natural formation processes that formed the site structures and middens. Using midden volume averaged from eight single hogan sites, and an ethnographic estimate of a maximum ten-year occupation span of a hogan or similar structure (Dean et al. n.d., Russell 1983; Weltfish 1965; Wilson 1934), Towner and Johnson derived an estimate of 1.05 cubic meters of trash accumulation per room per year. Although this method reduces the estimated occupation spans by half (Towner and Johnson 1998:178), it is still subject to the vagaries of accurate midden measurements and tends to ignore cultural processes, such as seasonality or periodic abandonment, that may seriously distort the estimates. Neither of these methods is appropriate for analysis of all the pueblito sites, primarily because the data concerning midden volume are not available. Therefore, I use the tree-ring data to estimate site occupation duration for those sites with appropriate tree-ring dates.

Figure 30 shows the difference between the earliest cutting or near-cutting date and latest date from all of the sampled pueblito sites. Fourteen of the sites did not yield cutting or near-cutting dates and cannot be used in the analysis. Differences between the types of dates at the remaining 50 sites range from zero years to more than one hundred years. A difference of "zero" years indicates that the earliest cutting date is in the same year as the latest date; commonly this situation indicates a single cutting date that is also the latest date from the site. There are situations, however, such as at Rincon Rockshelter and Hooded Fireplace, where a terminal cluster of cutting dates includes the earliest cutting date from the site. The majority of the sites (n=31) show a difference of ten years or less between the earliest cutting date and latest date; only seven sites, however, exhibit a difference of greater than 30 years between the earliest cutting and latest date. This distribution suggests that most of the sites were occupied for less than ten years.

Figure 31 shows the differences between the initial construction episodes as inferred from the tree-ring dates and the latest dates from all of the pueblito sites. Construction episodes could not be defined for 28 of the sites, and they are not included in this analysis. Differences between initial construction episodes and latest dates range from zero to 35 years. A large majority of the sites (n=30) show a difference of ten years or less, and at 17 of the sites there is no difference. Four of the five sites with differences of more than 10 years, Three Corn, Frances Canyon, Deer

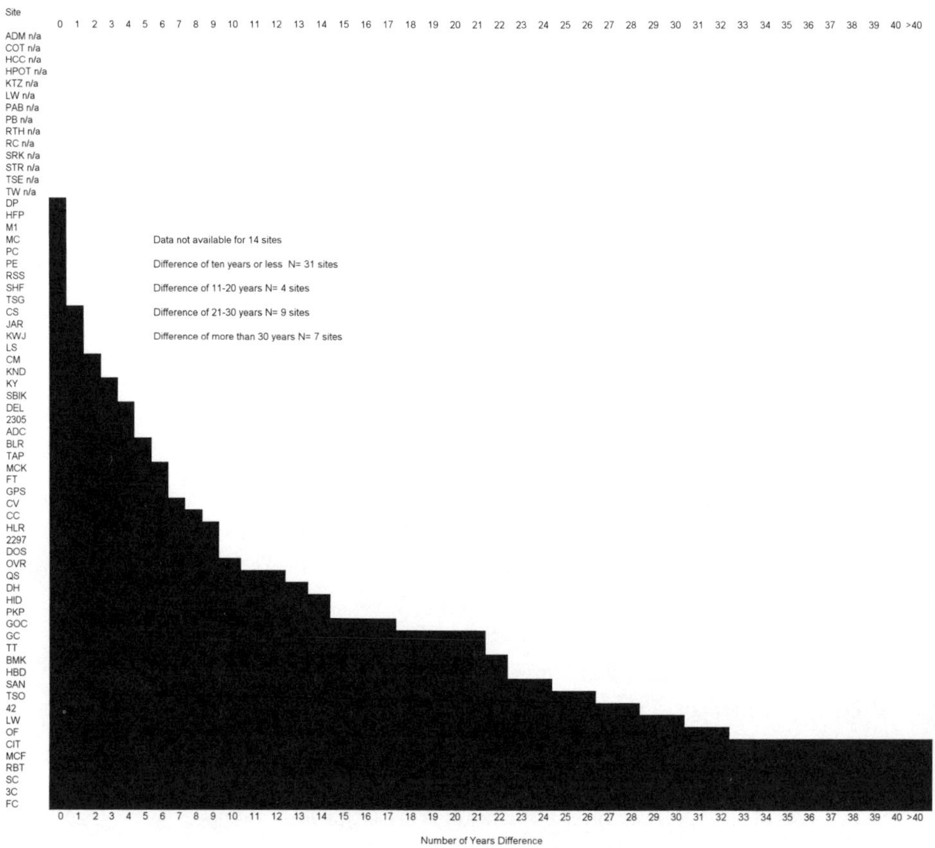

Figure 30. Graph of the difference between the earliest cutting or near-cutting date and the latest date from the sites

House, and Gomez Canyon, can be explained as the result of abandonment and reoccupation. At the other site, Buffalo Mask, the difference is the result of a late date from a notched log ladder that suggests the site was reused as a refuge but was never used as a habitation site.

Although the tree-ring data suggest the sites were occupied for only a short time, the possibility remains that (a) the structures were built and simply occupied without additional remodeling or repairs or (b) that such events left no tree-ring evidence. Complete sampling of all the datable timbers, combined with the low use-frequency of undatable species at the sites, makes the second possibility extremely unlikely. The first possibility, extended use of these structures without repairs or remodeling, is also unlikely. Ethnographic and archaeological evidence indicates that both wooden earth-covered structures and masonry or adobe structures have

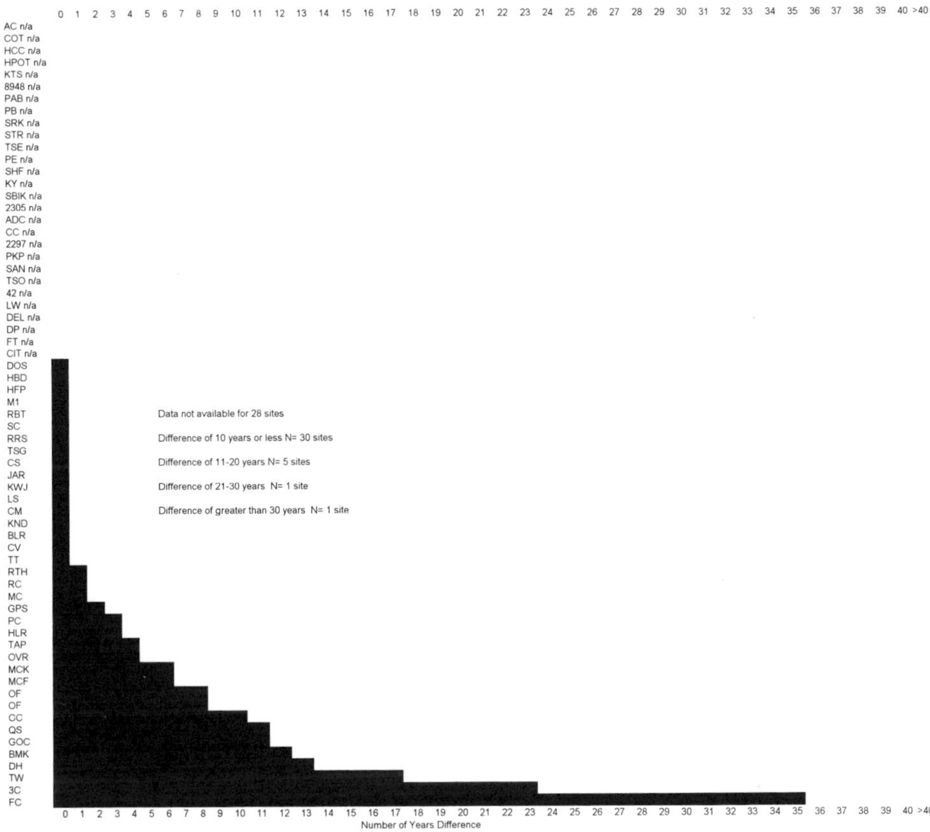

Figure 31. Graph of the difference between the earliest construction episode and the latest date from the sites

short use-lives, need frequent repairs, or both (McGuire and Schiffer 1983). Navajo hogans in other areas are typically occupied for periods of less than 8 to 10 years (Dean and Russell 1978; Oswald 1993; Rocek 1985, 1995; Russell 1983). They are often relocated and rebuilt because of insect infestation, but rebuilding normally introduced new wooden elements into the structure and would alter the tree-ring date distribution. Similarly, Hidatsa earthlodges were usually occupied for 10 years or less (Wilson 1934), and Pawnee earthlodges for a maximum of 15 years (Weltfish 1965). Ahlstrom (1985) estimates that Basketmaker III pit houses were used for less than a decade, but often required repairs during their use-lives.

Masonry structures may have had longer use-lives because the weight-bearing beams were not in contact with the ground and were not subject to rot or deterioration. They still may have become infested with insects, however, and undoubtedly required frequent maintenance. Adobe or mud-brick structures in the Middle East require annual or even more frequent maintenance, usually in the form of mud plastering, but occasionally by replacing wooden structural members as well (Bultnick 1969:69; Carter and Pagniero 1966; Moorey 1994). Rothschild et al. (1993:126) indicate that in the Zuni area, if a structure is not maintained at least every 6 to 12 months, "sections of adobe or stone masonry move or break off, doors and window frames shift, and roofs collapse." They also show the substantial decline of a masonry building in Upper Nutria over the course of a decade (Rothschild et al. 1993: Figs. 10.12 and 10.13). An additional, and in some ways more interesting, line of evidence on the rate of beam replacement in masonry structures is an example from historical archaeology. Baxter et al. (1997) document a major reroofing episode at the reconstructed Kinishba Ruin in 1950. Byron Cummings (1940) reconstructed the ruin beginning in 1933 and continued until the mid-1940s. The reconstruction documentation (Cummings 1940; Reid and Whittlesey 1989) indicates that repairs to the reconstructed pueblo and repair of some rebuilt roofs began less than a decade after the original reconstruction. Tree-ring dating of the reconstructed pueblo (Baxter et al. 1997) indicates that even these repair roofs were completely rebuilt less than five years after they were repaired. Finally, an experimental pit house with creosote-treated pecan roof beams constructed by Pima College in Tucson is beginning to show signs of roof decay after only five years (D. Stephens, personal communication 1996). Thus, the rate of repairs on poorly maintained masonry structures, including roof beam replacement, is often quite rapid and usually less than five years. The ethnographic, archaeological, experimental, and tree-ring data indicate that most, if not all, masonry structures require beam replacement in fewer than ten years after initial construction. I infer, therefore, that most pueblito sites were occupied for less than ten years, and many were probably used for fewer than five years. As discussed in Chapter 6, such short occupation spans have serious implications for questions concerning the size and mobility of the eighteenth-century Dinétah Navajo population.

Regional Occupation Density

If not all pueblito sites were occupied throughout the Gobernador Phase, it is important to estimate when how many sites were occupied simultane-

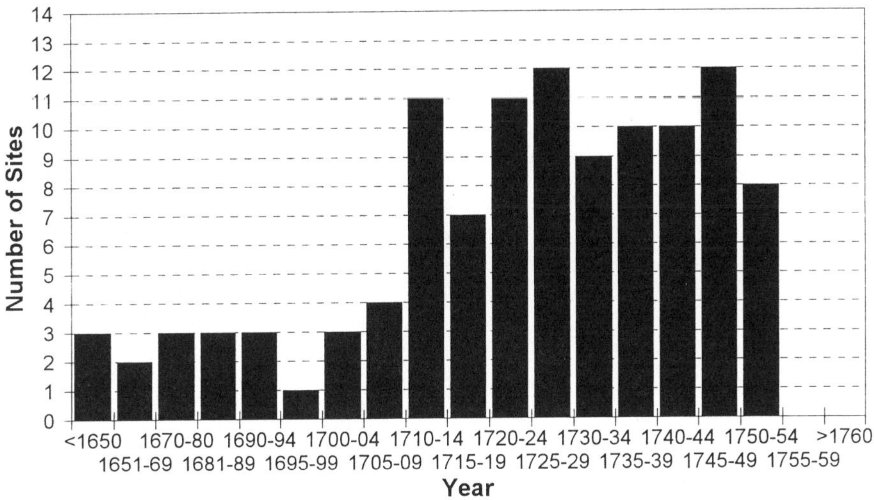

Figure 32. Frequency of sites with cutting or near-cutting dates per five-year interval

ously. Because cutting dates and defined construction episodes are not available for every site, I must use a combination of factors to estimate site occupations. The tree-ring data, combined with the defined construction episodes and occupation duration estimates, provide a powerful tool for determining the occupational intensity of the Dinétah during the eighteenth century. It is important to remember, however, that the estimates are based only on the tree-ring data from the pueblito sites.

As with occupation duration, there are several methods for estimating occupational intensity. Marshall (1991, 1995) has shown that the pueblitos are parts of much larger communities; the previous section of this chapter suggests that the sampled hogans are contemporary with the pueblitos. Towner and Johnson (1998) and Dykeman and Wharton (1996) demonstrated that not only are there hogan communities in the immediate vicinity of pueblitos, but that there are many other hogans scattered in the surrounding areas. The most detailed analysis of eighteenth-century hogan communities has been conducted by Sesler, Hovezak, and Wilshusen (2000) on Frances Mesa. They clearly domonstrate that hogans were dismantled and rebuilt within communities and that communities were relocated on the landscape with some frequency. I suggest that most pueblitos were part of the same relatively mobile settlement pattern, but that they were not dismantled, simply disused. Thus, the data generated from the pueblitos should not and cannot be used to extrapolate population figures

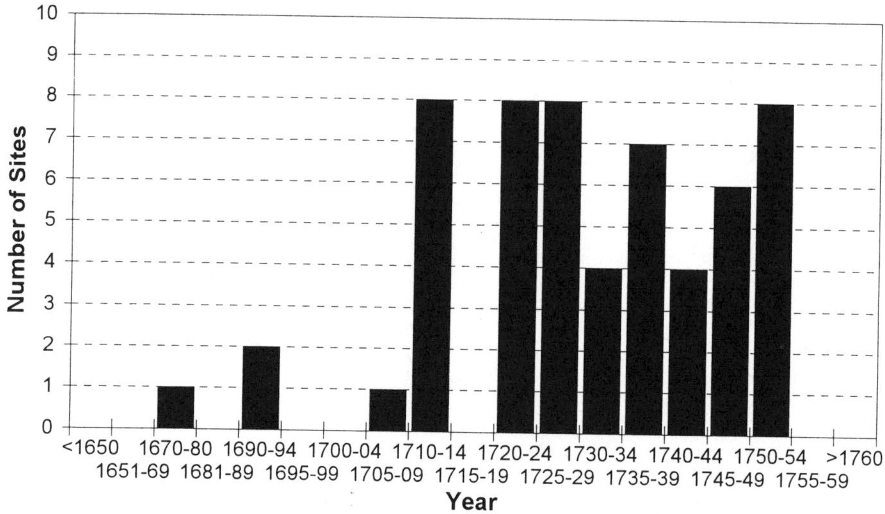

Figure 33. Frequency of sites with identified construction episodes per five-year interval

across the entire Dinétah; they can be used, however, to show the intensity and episodic nature of the pueblito site occupations.

Figure 32 shows the frequency of sites with cutting or near-cutting dates in each five-year interval during the pueblito site occupation. Two aspects of Figure 32 warrant attention. First, the data indicate that there was little construction activity in the area prior to 1710 and the activity that did occur was confined to a few sites. This low level of activity is reduced even further if most pre-1710 cutting and near-cutting dates are considered to be examples of dead-wood use or beam reuse. Second, between 1710 and 1754, the number of sites with evidence of tree-harvesting remains relatively constant. These data suggest that only between seven and twelve pueblito sites were ever occupied at the same time.

The intensity of pueblito site occupations is much lower if only those sites with defined construction episodes are considered (Figure 33). Prior to 1710, only Buffalo Mask (1680), Tapacito (1690–1694), and McKean Pueblito (1708) show any construction activity; Buffalo Mask and Tapacito, however, were probably abandoned before A.D. 1710. The major period of the pueblito occupation, 1710–1754, shows a fluctuating number of pueblitos undergoing construction, but there is no evidence of building at more than eight sites. Although the number of simultaneously occupied sites is generally small, it may be even more significant that only rarely

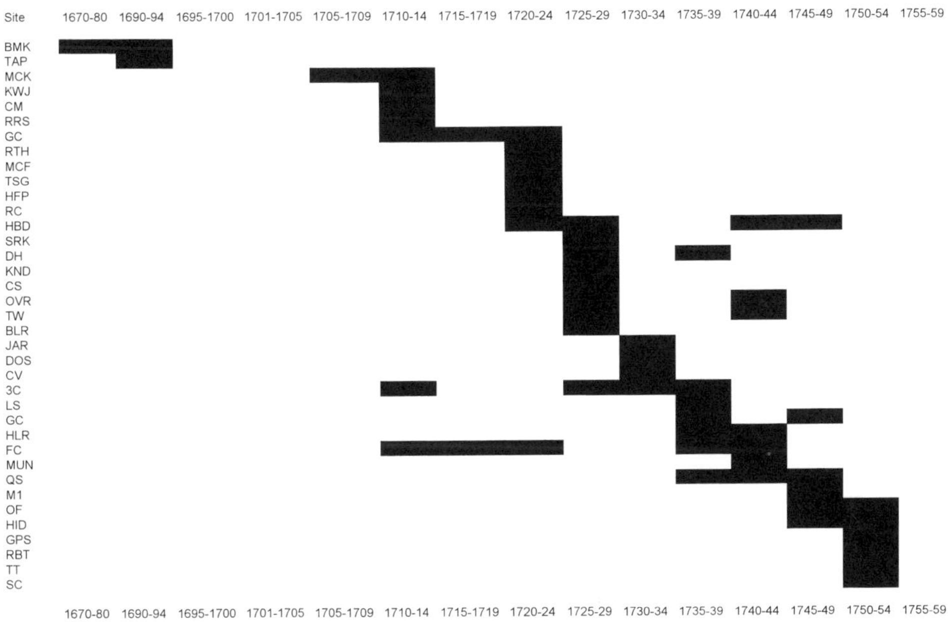

Figure 34. Temporal distribution of site construction and later occupation events

does a site contribute to sequential intervals; that is, few sites show evidence of construction in consecutive five-year periods.

Of the 36 sites where construction or later occupation events can be identified, those events occurred within a 5-year interval at 25 of the sites. Not surprisingly, it is the larger sites that show evidence of construction over longer periods of time (Figure 34). Frances Canyon was apparently used the longest, with building activities in the 1710–1714, 1720–1724, 1735–1740, and 1740–1745 intervals; Three Corn Ruin, exhibits building in the 1710–1714, 1725–1729, 1730–1734, and 1735–1739 intervals; and Garcia Canyon, shows possible building in 1710–1714, 1715–1719, and 1720–1724. McKean Pueblito is somewhat unusual in that it shows construction in the 1705–1709 and 1710–1714 periods, with the later dates probably indicating repair, not new construction. It should be remembered, however, that both Frances Canyon Ruin and Three Corn Ruin were probably abandoned for 5 to 10 years between occupation events and that the proposed construction sequence at Garcia Canyon Pueblito is based on major rebuilding in 1722. Several sites, including Twine House (1728–1745), 42 Pueblito (1727–1741), The Overlook Site

(1727–1741), Deer House (1726–1730 and 1735–1739), Gomez Canyon (1735–1739 and 1745–1749) show evidence of building in two nonsequential intervals, but all may have been abandoned between construction events. Three sites show continuous occupation in two consecutive intervals. Hill Road Ruin exhibits construction in the 1735–1739 and 1740–1745 intervals, but earlier construction was done in 1738, and the later construction was completed by 1741; thus, the site spans two intervals but was built in a three-year period. Old Fort Ruin was apparently continuously occupied for at least eight years (1746–1754), and Hidden House (1746–1754) may have been part of the same occupation of San Rafael Canyon (Towner and Johnson 1998).

Spatial Aspects of the Pueblito Site Occupations

If the pueblito sites were not occupied for long periods of time, and I can identify which sites were occupied during which temporal intervals, the next step in analyzing their distribution is to place those occupations in their proper geographic positions. This is one approach to (a) determining if there were settlement shifts within the Dinétah during the first half of the eighteenth century, and (b) identifying possible "pueblito communities" that extend beyond the boundaries of individual sites. Although there is no doubt that undiscovered and undocumented pueblito sites exist in the area, the current sample size probably represents a fairly accurate description of their geographic and temporal distributions.

Tree-ring-dated pueblito sites are spread throughout the Dinétah, both north and south of the San Juan River (Figure 4). The Largo drainage and its tributaries contain the most pueblitos and the smaller, more northerly drainages contain relatively fewer pueblito sites. There are no known, or at least dated, pueblito sites north of the La Jara drainage. But I believe this lack of evidence to be an artifact of survey and sample collection strategies, not necessarily a reflection of eighteenth-century Navajo settlement dynamics. Only three dated sites (Pueblito Canyon Pueblito, Pueblito East, and the Cabresto Mesa Tower Complex) are present in the La Jara drainage, and all are near where it widens after descending Vaqueros Canyon. Frances Canyon, the next San Juan tributary south of the La Jara, contains only two dated sites, Frances Canyon and Garcia Canyon, although at least two other pueblitos are present as well (Powers and Johnson 1988). Finally, the last northern drainage, Gobernador Canyon, contains fourteen dated pueblito sites and many more undated ones (Enloe et al. 1973; Powers and Johnson 1987; Towner and Johnson 1996). Interestingly, however, pueblito sites tend not to be located near the confluence of the San Juan and Los Piños, one of the most sacred places in Navajo religion (Van Valkenburgh 1974).

The Largo drainage and its tributaries contain 46 dated pueblito sites and many more undated ones (Powers and Johnson 1987; Copeland, personal communication, 1996). Only 10 sites, however, are located in the Largo drainage proper; the remainder are in tributary canyons or on mesa tops away from the drainage. Relative to their sizes, smaller tributaries such as Crow and Palluche Canyons, hold most of the pueblito sites. The two main tributaries of the Largo, Cereza Canyon, and Blanco Canyon, are interesting for their contrast. Cereza Canyon and its tributaries contain 15 dated pueblito sites, but not a single dated (or undated) pueblito is known from Blanco Canyon. This difference is probably related to the different topography of the two areas. Whereas Cereza Canyon cuts sedimentary rocks to form steep-sided canyons with flat floodplains, Blanco Canyon is on the edge of the San Juan Basin and, especially on its western side, is composed of relatively flat rolling hills. The latter area is certainly not appropriate for building pueblitos, although other eighteenth-century Navajo site types may be located there (Gilpin 1993a, 1993b).

The Cereza drainage, like the main Largo drainage, contains only a few pueblito sites, and only two of those produced dates. Again it is the tributaries, such as Jesus, Delgadito, Adams, and Muñoz Canyons that contain most of the sites. Only two dated sites, Hill Road Ruin and Kin Wólá Jíní, are located in the main Cereza Canyon, although several others overlook it. Interestingly, there appears to be a high frequency of pueblitos on Ensenada Mesa between the two large canyons, particularly in the areas of Smith, Crow, Adolpho, and Gomez Canyons. Whether this is simply a function of topography or some other natural or cultural phenomenon remains unknown.

I have suggested elsewhere (Towner 1996, 1997) that there were geographic shifts in the pace of pueblito construction in the Dinétah during the early and mid-1700s. I believe the evidence is clear that Tapacito Ruin was an isolated masonry structure during the entirety of its occupation (Towner 1997; Towner and Dean 1992). The tree-ring data show that Buffalo Mask may also be unique. These are the only pre-1700 pueblito sites, and they are located in very different geographic and topographic settings. The earliest site, Buffalo Mask, is located in a small alcove high above a small tributary of Jesus Canyon. The site is hidden, and the paucity of occupation debris and lack of hogans nearby suggest that it served only as a refuge in emergency situations, or possibly as a ceremonial site. There is historical evidence that Los Peñoles (LA 8948) was used as a refuge in 1705 but that it probably was not a habitation (Hendricks and Wilson 1996). Tapacito Ruin, on the other hand, was part of a larger community on a broad bench above the largest drainage in the Dinétah. Although it is somewhat hidden, it has no apparent defensive features and lacks the spectacular setting of most other pueblitos.

Pueblito Temporal Variability

All of the pueblito sites were occupied over a relatively brief 70-year period from about A.D. 1680 to A.D. 1760, but they vary tremendously in their temporal and spatial placement, as well as in their architectural attributes and artifact assemblages. It is probably no coincidence that historical records and Navajo oral traditions indicate tremendous political, cultural, and environmental perturbations at the same time.

Powers and Johnson's (1987) five-part classification system for pueblitos suggested that there was an evolution of pueblito styles related to the adaptation of Puebloan refugees. In short, they suggested that the earlier sites should show the most Puebloan attributes, and the later sites should show more adaptation to local physical and social environments. Although subsequent tree-ring dating of many sites showed their scheme to be temporally flawed, the concept of pueblito variability in response to differing social conditions is still valid.

Early Pueblito Sites (A.D. 1680–1720). There are 18 known pueblito sites that were built prior to A.D. 1720 (Figure 35), including four small mesa rim sites, four large, multi-room open sites with expansive viewsheds, two defensively oriented forked-pole hogan sites, six rockshelter sites, Los Peñoles, and Tapacito Ruin. As discussed previously, Tapacito and Buffalo Mask should both be considered separately from pueblito sites.

The mesa rim sites include Kin Wólá Jíní, built in A.D. 1712 and probably occupied only for a few years; Garcia Canyon Pueblito, built in A.D. 1712 and extensively remodeled in 1722; Pueblito East Ruin, probably built in 1713; and McKean Pueblito, built in 1708 and repaired in 1713.

The four large multi-room open structures built prior to A.D. 1720 include Cabresto Mesa Tower, built in 1713–1714; Ridgetop House, constructed sometime about 1720; Frances Canyon Ruin, built between 1710

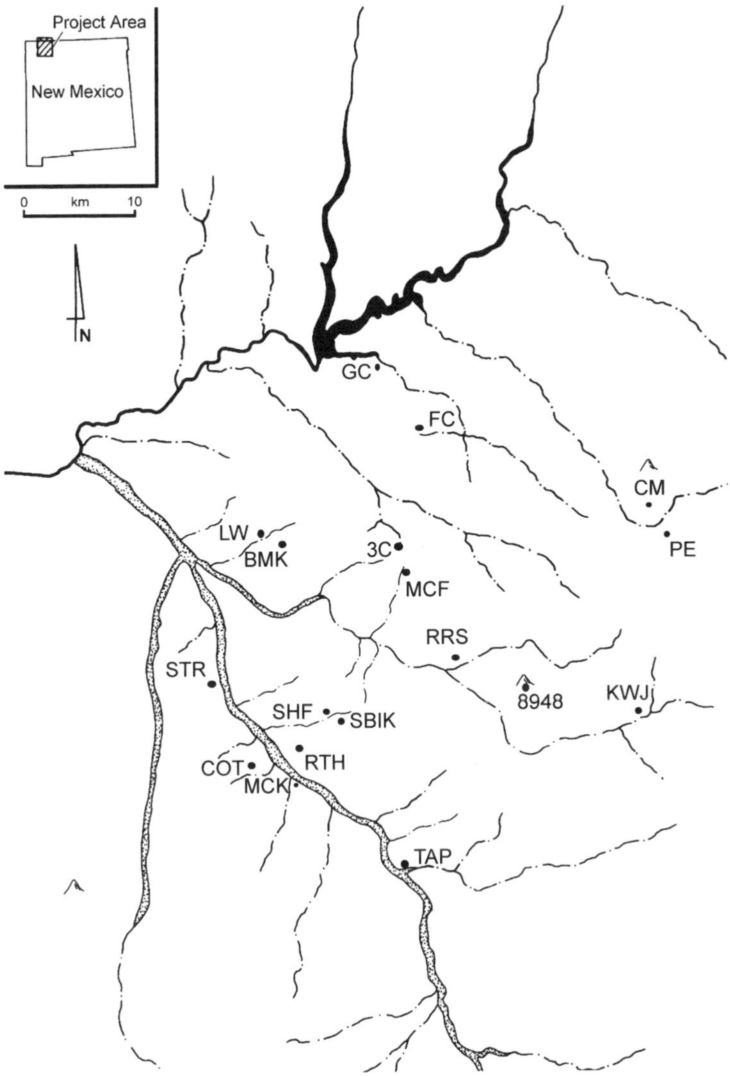

Figure 35. Pueblito sites built prior to 1720

and 1745 (most construction was completed by 1722); and Three Corn Ruin, built between 1714 and 1734 (again, most construction was completed by the 1720s). Frances Canyon Ruin and Three Corn Ruin eventually became two of the largest Navajo-built structures in the Dinétah. By about 1720, these sites contained at least 27 masonry rooms and possibly a score or more hogans. The two defensively oriented hogan sites that were probably occupied prior to 1720 include Star Rock Refuge built after 1708 and Cottonwood Divide occupied after 1715. Discounting Buffalo Mask, the six rockshelter sites occupied prior to 1720 include Morris Cliff Dwelling, built in 1720; Rincon Rockshelter, built in 1713; Shándin Bíikin, occupied after 1705; Shaft House, built after 1712; and Lone Warrior Pueblito, built sometime after 1682. Dating of these sites is less certain than for the large masonry ruins, but all appear to have been constructed after 1705 and prior to about 1720. All of these sites are located in small side canyons away from large drainages in the central Largo area.

There are several interesting aspects to all the early nonrockshelter sites, and especially the large sites. The large sites are located on prominences, and all are intervisible (Jacobson et al. 1992), although all are located more than 16 km (10 miles) from each other. The defensively oriented hogan sites are located on isolated prominences adjacent to Blanco Mesa and may be visually connected to the more easterly sites. One interpretation of this distribution is that the large sites, with the exception of Kin Wólá Jíní, comprised substantial Navajo communities that coalesced in response to the Spaniards' entradas into the Dinétah. It is the size, and especially the locations, of these sites that suggest they were built to counter the Spaniards' efforts. Although large, it is unlikely that the Navajo population could have withstood a direct confrontation with the vastly superior Spanish forces. By locating their sites on inaccessible prominences, however, the Navajos somewhat negated the Spaniards' numerical and technological advantages. This strategy had proven effective at Los Peñoles in 1705, and may have been subsequently used by some Navajo groups. It is important to note that most of the early nonrockshelter sites are intervisible (Jacobson et al. 1992). Significantly, Cabresto Mesa Tower and Pueblito East seem designed to "guard" or "monitor" Vaqueros Canyon, the route into the Dinétah (see below) used by Roque de Madrid and presumably later entradas. The intervisibility of the other pueblitos may have provided those Navajo populations with significant advance warning of Spanish invasions.

The rockshelter sites, most of which are associated with forked-pole hogans, were probably occupied by single or extended family groups. Located in small side canyons, all of the rockshelter sites may represent a "stay hidden strategy." Such a strategy may have been effective because the Spanish entradas, with their large numbers of Indian auxillaries, livestock, and material, only traversed the Dinétah's major drainages and mesa tops (Hendricks and Wilson 1996). Apparently, both the "be inaccessible" and "stay hidden" stategies were successful, because few direct confrontations with the Spaniards occurred, and they ceased invading the area in the mid-1710s.

Middle Period Pueblito Sites (A.D. 1720-1745). Tree-ring-dated pueblito sites built between about 1720 and 1745 (n=36) are architecturally and locationally different from the earlier sites and comprise the largest number of pueblitos built during any single time period (Figure 36). There is no tree-ring evidence of construction in rockshelters or the initial construction of large multi-room open sites during this era. Although Three Corn Ruin, Frances Canyon Ruin, and Garcia Canyon Pueblito show some signs of continued occupation, Towner (1997) suggests that all may have been periodically abandoned. Probably, most of the earlier sites were not occupied much later than 1720. The majority of the pueblitos built in the 1720–1745 time period are small—one- to five-room structures built on boulder tops or mesa rims. All are associated with contemporary forked-pole hogans and other Navajo features.

A good example of this type of pueblito is Twine House (Ababneh et al. 2000). Located atop a boulder on the floor of Palluche Canyon, Twine House was built in 1727. It is spatially associated with several other pueblitos and forked-pole hogan complexes in the canyon, but is not part of a visually connected network of sites. It contains only three rooms and occupies the entirety of the boulder top, with an off-set entry and off-set interior doorways. Thus, Twine House appears to be an "individually defensive" site.

Historical documents and archaeological remains suggest that by 1727 the San Juan River may have been a temporary physical barrier between the Navajos of Dinétah and the Utes of southwestern Colorado (Schaafsma 1996). The historical record also suggests that Ute raiding of both Spanish and Navajo settlements had become a serious problem by 1716 (Hill 1940; Reeve 1959), probably due to increased Ute access to Spanish horses and French guns (John 1975). Indeed, the Spaniards' "peace" with the Navajos lasted from approximately 1716 to 1770. This peace left the Navajo to defend themselves against more mobile and better armed Ute and Comanche attackers.

Unlike the Spaniards' objectives during the earlier entradas of "punishing" the Navajos or retrieving "captives," Ute raiding goals appear to have been oriented toward material gain. The Navajo at this time had a

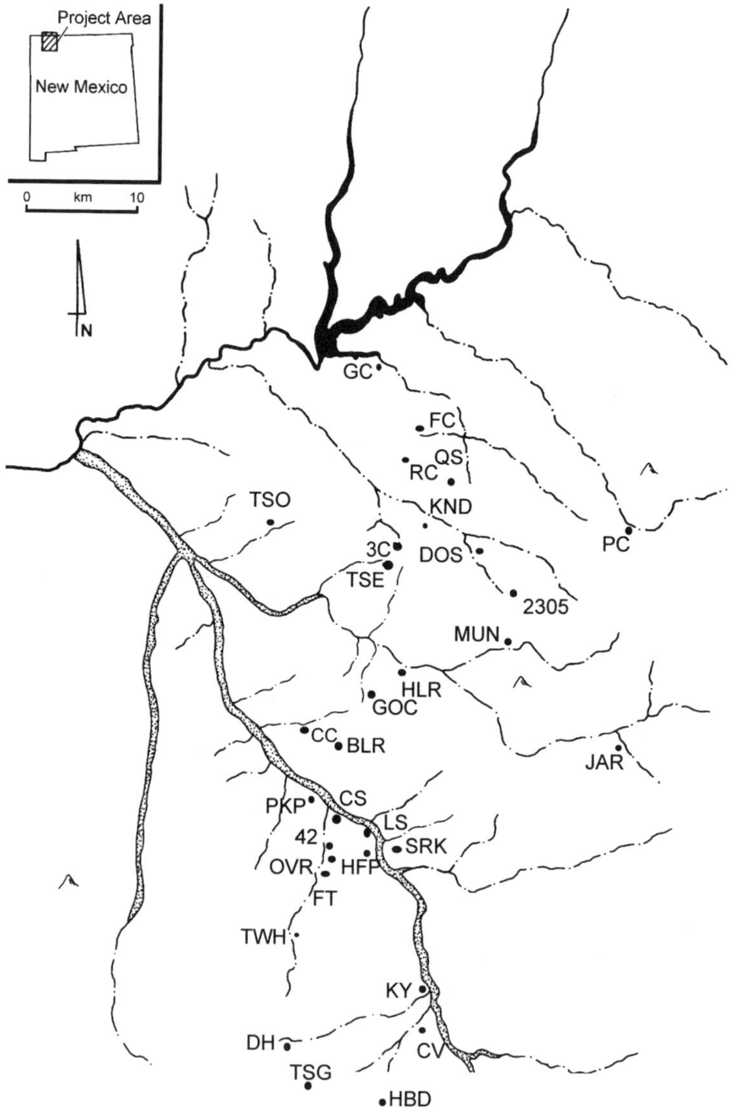

Figure 36. Pueblito sites built between 1720 and 1745

hunter-gatherer–agriculturalist-pastoralist economy and were considered relatively wealthy. Raiding for subsistence items, such as corn, hides, sheep, or pinyon nuts, however, was probably not a worthwhile effort. Foodstuffs, such as corn or nuts, would have been difficult to gather and bulky to transport, especially over the San Juan River with Navajo pursuers. Although the Ute probably were able to acquire some horses on raids, the idea of acquiring a substantial number of sheep on such a raid evokes comical images of smoke, gunfire, bleating sheep, and mass confusion—hardly an ideal raiding scenario.

There was one "commodity," however, that could be procured quickly, transported easily, and which was very valuable—human beings. Although slavery was officially illegal in eighteenth-century New Spain, non-Christian Indians could be enslaved in order to convert them to the faith and thus "save" their souls (Brugge 1985). This legal loophole created a thriving slave market in northen New Spain. Women, and especially children, often ended up as household servants in the haciendas of the Rio Grande Valley; whereas the more dangerous adult males were transported to the mines of northern Mexico. The value of such human commodities was high indeed; in 1711, a single Navajo child brought three-to-five horses at the trade fairs in Taos and Pecos (Kessell 1987:364). Given the Ute desire for horses and the higher risk of capturing and transporting adults of either sex, it is probable that the primary objective of most Ute raids was to capture Navajo children to be sold into slavery, and secondarily, to procure Navajo horses.

The defense against such raids was simple, but effective. By building small pueblitos near a cluster of forked-pole hogans, the Navajo had a more secure temporary refuge when danger threatened. If a threat was detected, the site occupants could retreat to the higher ground of the pueblito. If the raiders continued to attack, they were forced to dismount and fight their way "uphill" into a structure that was architecturally designed to be defended by few occupants. Thus, the cost of such a raid was raised dramatically in terms of lost speed, increased danger to the attackers, and lessened probability of securing captives or horses. Mesa rim pueblitos functioned in a similar manner, by forcing the raiders to dismount and attack from below or dismount and attack a well-defended, architecturally complex entrance. In either case, the raiders could not ride through a settlement and accomplish their goals without substantially more risk and violence.

Although the large sites may have still been used by this time, the archaeolgical record suggests that rockshelters were not. The archaeological and dendrochronological records further indicate that most pueblito sites built between about 1720 and 1745 were occupied by single- or extended-family Navajo groups responsible for their own safety. These sites were not, however, isolated refuges on the landscape. Most were probably parts of local communities (see below) that may represent eighteenth-century land-tenure systems. Once built, such sites may have become part of the "cultural geography" of an area, even if they fell into disuse (e.g., Basso 1996).

Post-1745 Pueblito Sites. Pueblito sites constructed after 1745 (Figure 37) are more variable than those built during the earlier periods, but do seem to fit an important pattern. The most significant of these sites is Old Fort Ruin. Old Fort was constructed on the first bench above San Rafael Canyon (Towner and Johnson 1998), and includes 15 masonry rooms, six forked-pole hogans, and a sweatlodge. Importantly, the entire site complex is enclosed by a massive two-meter-high wall that was built in 1750. Two securely dated contemporary boulder-top pueblito and hogan complexes (Hidden House and Rabbit House) are located across the canyon, and both are clearly visible from Old Fort. Towner and Johnson (1998) suggest that the settlement pattern in San Rafael Canyon changed from the early 1714–1730 period centered around Three Corn Ruin and its long-distance visibility to a 1745–1755 occupation centered around Old Fort and its nearby satellite sites with only intracanyon connections.

The other post-1745 pueblitos include Morris 1, built in 1748–1749; Gould Pass Puelbito, built in 1752; Truby's Tower, built in 1753; and Simon Canyon, built in 1754. Morris 1 was part of a community that had occupied Romine Canyon since the early 1600s (Dykeman and Wharton 1995), and resembles the earlier boulder-top pueblito configuration. Simon Canyon, a single-room structure atop a fifteen-meter-high boulder, is unique in three ways. It is the latest known pueblito, it is the only dated pueblito not associated with forked-pole hogans or other Navajo features, and it is the best-dated pueblito north of the San Juan River (see below for a discussion of other Navajo sites north of the San Juan). Truby's Tower is a seemingly isolated boulder-top structure high above Largo Canyon. Because no large surveys (cf. Dykeman and Wharton 1994, 1996; Marshall 1991; Towner and Johnson 1998) have been conducted around Truby's Tower, it is possible that contemporary forked-pole hogans did exist in the area. Gould Pass Pueblito is a two-room structure associated with two forked-pole hogans. It is located on a small outcrop on top of Ensenada Mesa and has an extensive viewshed to the south, west, and northeast.

Despite their variability, the post-1745 pueblitos may fit a pattern similar to that of the early pueblitos. Old Fort was clearly a major population center and may have been the focus of the San Rafael community in the

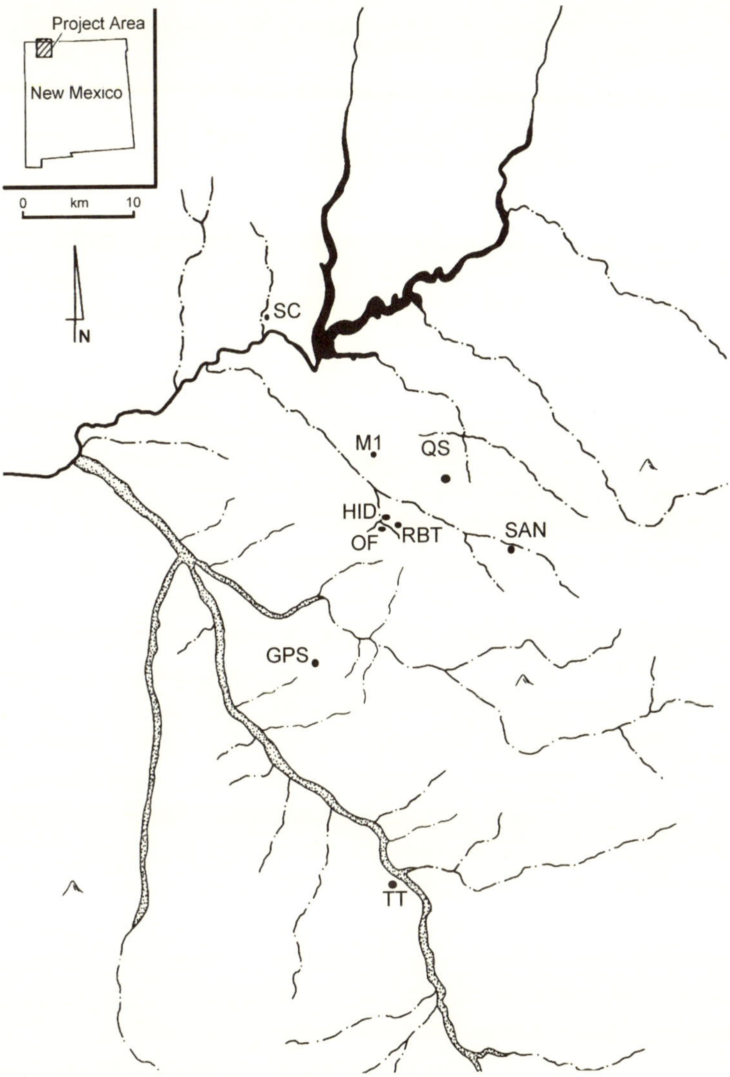

Figure 37. Pueblito sites built between 1745 and 1755

late 1740s and early 1750s (Towner and Johnson 1998). This community, with its "walled town," may not have needed long-range visual contacts for protection against small raiding parties. The other late sites, with the exception of Simon Canyon, all have extensive viewsheds in one or more directions. Morris 1 is located below the rim of Frances Mesa, and has extensive visibility only to the southern horizon—possibly to some of the sites in San Rafael Canyon (but not Old Fort). It may be significant that the northern approach to Morris 1 is across Frances Mesa—the location of a major Navajo occupation during the 1740s (Sesler, Hovezak, and Wilshusen 2000). Truby's Tower commands a view at least 15 miles (24 km) up and down Largo Canyon, and into many side canyons as well. As mentioned above, Gould Pass Pueblito commands a view in almost all directions—at least 20 miles (30 km) to the south, 10 miles (16 km) to the west, and 3 miles (5 km) down Martin Canyon to the northeast. It may also be significant that an undated pueblito, LA 21712, is located at the bottom of the north side of Gould Pass. The only late pueblito without an expansive viewshed is Simon Canyon Pueblito, a structure located on top of a large boulder, but in the bottom of a steep-sided canyon. Any explanations of Simon Canyon are speculative, but perhaps it was built specifically to monitor Ute activities north of the river, or in response to an emergency situation (i.e., Navajos trapped in Ute territory).

With the exception of Old Fort and Simon Canyon, the late pueblitos seem to combine the long-distance visibility of the early sites with the smaller size of the middle-period pueblitos. This pattern suggests a different defensive strategy, probably in response to a different threat.

The historical record, although scanty, provides hints of the nature of warfare in the northern Southwest in the late 1740s. Schroeder (1965) indicates that in 1747 the Utes and Comanches staged a major attack on the Spanish town of Abiquiu, after which the town was abandoned (Twitchell 1914). Indeed, several Spanish settlements in the north were "in ruinous condition" during the 1740s and 1750s (Schroeder 1965:59).

In 1752 the Utes reportedly attacked the Navajo mesas "with such force that the Navajos found the action bloody for them; some were killed, others captured, with no danger to the Utes who strove for a complete victory by closing in to reach the tops of the mesas" (Reeve 1960:201). By 1754, Tomas Velez Cachupin, the governor of New Mexico, indicated that the Navajos had abandoned Largo Canyon, "fleeing the war by which the Utahs seek satisfaction for injuries done to them by the Navajo . . . the Utahs have so frequently made war on the Navajos and punished them that they have caused them to flee from the province" (Reeve 1960:202).

The late pueblitos, therefore, seem to have been built to counter quick-strike raids, but also to protect larger numbers of people from more sustained and serious violence. Long-distance communication may have offered some advanced warning of threats, and substantial, well-organized local communities may have facilitated responses to Ute incursions. Ultimately, however, the pueblito system failed and the Dinétah Navajo migrated south toward Mt. Taylor and west toward the Chuska Mountains.

There are no tree-ring-dated Dinétah pueblito sites that postdate 1754, and very few tree-ring dates from any Dinétah Navajo sites. A noncutting date of 1756++vv from a scarred tree near Old Fort Ruin is the latest date associated with a pueblito (Towner and Johnson 1998); Sesler and Hovezak (1996) report a 1770+vv date from a ramada at LA 79462, although they suggest the wood may be from a historic structure. Thus, the tree-ring data indicate that the Navajo occupation of the Dinétah ended prior to about A.D. 1760 (but see below regarding data from the Aztec area).

Identifying Interpueblito Navajo Communities

The pueblito sites have been described as "site complexes" (Marshall 1991) because of the presence of forked-pole hogans, masonry structures, sweatlodges, and other structure types. Certainly, most forked-pole hogans are contemporaneous with the masonry structures and, as Mashall correctly asserts, they formed substantial communities. The interaction of these "site-based" communities with other Navajo communities in the area is a topic that may provide insight into the organization of Dinétah Navajo culture on a wider spatial scale.

Jacobson et al. (1992) used a computer-based viewshed analysis to suggest that the pueblito communities were integrated into a broader regional network of visually connected sites. Unfortunately, their use of 1:250,000-scale maps and reliance on locational data in the New Mexico ARMS files often yielded erroneous results. For example, they show that Tapacito and Split Rock Ruins are intervisible; however, my field testing indicates that they are not. In addition, many of the sites were not contemporaneous; if the sites were not occupied simultaneously, it is irrelevant if they are intervisible. It is important, therefore, to determine which sites were occupied simultaneously and could have been interconnected.

The pueblito communities described below (Figure 38) are defined on several criteria. First and foremost, sites must have been simultaneously occupied to be considered part of the same community. Tree-ring data are used to assess contemporaneity, but noncutting dates are used much more liberally than in the above analyses. Therefore, my delineation of these communities should be considered tentative and should be tested against additional data. Second, the pueblito sites must be spatially related to be

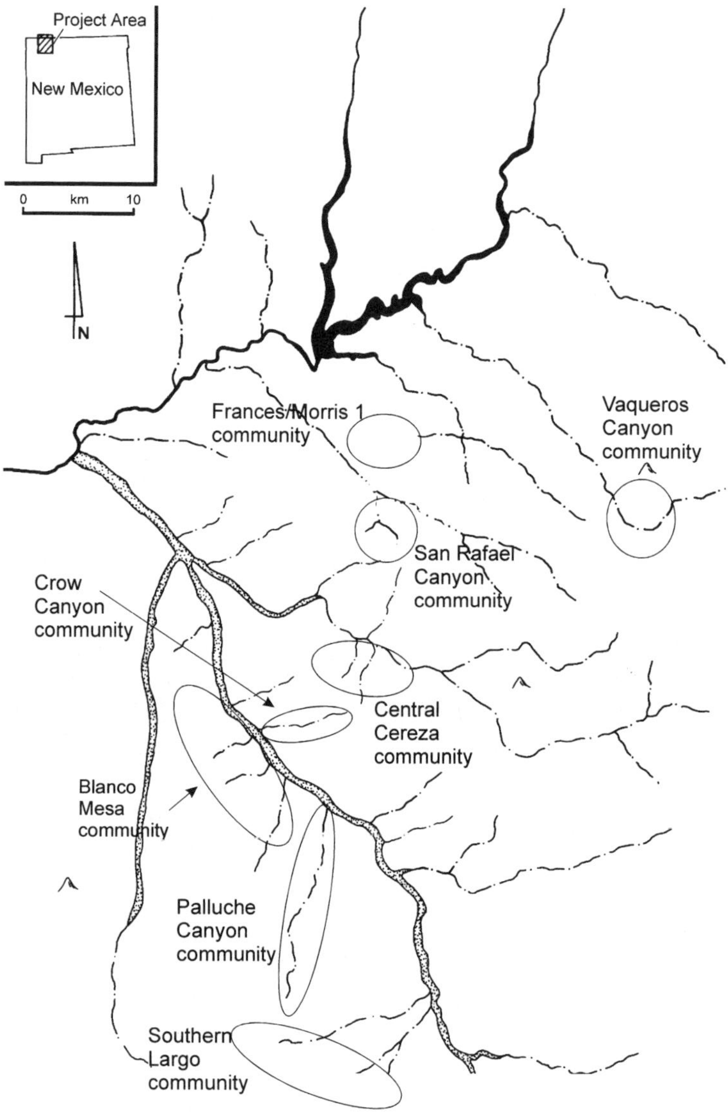

Figure 38. Interpueblito Communities in the Dinétah

considered part of the same community. Spatial relationships, however, are considered at different geographic scales. At a "local" scale, intervisible contemporaneous pueblitos and undated intervisible pueblitos with similar topographic and architectural attributes are considered parts of the same community. At a much larger, "extralocal" scale, I hypothesize long-distance connections between individual sites—and perhaps local communities—based on roughly contemporaneous tree-ring dates and intervisibility. There is one caveat to this "community" analysis. Sesler, Hovezak, and Wilshusen (2000) have demonstrated the existence of eighteenth-century forked-pole hogan communities not associated with pueblitos in the Dinétah. Clearly, pueblitos are only one component of the Dinétah Navajo occupation of the area and my hypothesized pueblito communities should be considered in a more fine-grained context. Until large-scale intensive surveys, like the Frances Mesa project (Sesler, Hovezak, and Wilshusen 2000) are conducted in other areas, however, these pueblito data offer the only clues about the use of the landscape. But they should not be considered definitive.

The San Rafael Canyon Community. Towner and Johnson (1998) have discussed two different communities in San Rafael Canyon. The earlier community, occupied during the 1710s and 1720s, was focused on Three Corn Ruin. The extensive middens around the pueblito attest to a substantial occupation, as do several dated forked-pole hogans. Other pueblitos, all undated, that may have been part of this community are located high above the canyon floor include LA 11089 (Morris Site 2), LA 11153, and LA 11187. All of these structures are intervisible and could have been part of the same communication network.

A later community apparently developed at lower elevations in San Rafael Canyon after Three Corn was abandoned. By the late 1740s, Old Fort Ruin was the focus of the settlement. Other dated pueblitos that were occupied at the same time, and which are intervisible, include Hidden House and Rabbit House on the north side of the canyon. Undated pueblitos that may have been part of the same community include Kin Tsézéí, LA 11146, LA 11161, and LA 11171, all located on the first bench above the canyon floor (Figure 39).

Reasons for the shift in the community remain unclear. Towner and Johnson (1998) suggest that the later (Old Fort) community maintained fewer long-distance connections than the earlier (Three Corn) community. Such a hypothesis may be related to changes in the nature of warfare during the first half of the 1700s (Towner, Rogers, and Copeland 2001) and is discussed below.

The Vaqueros Canyon Community. Vaqueros Canyon enters the northeastern Dinétah from the highlands near *Ch'óol'í'í* (Gobernador Knob)

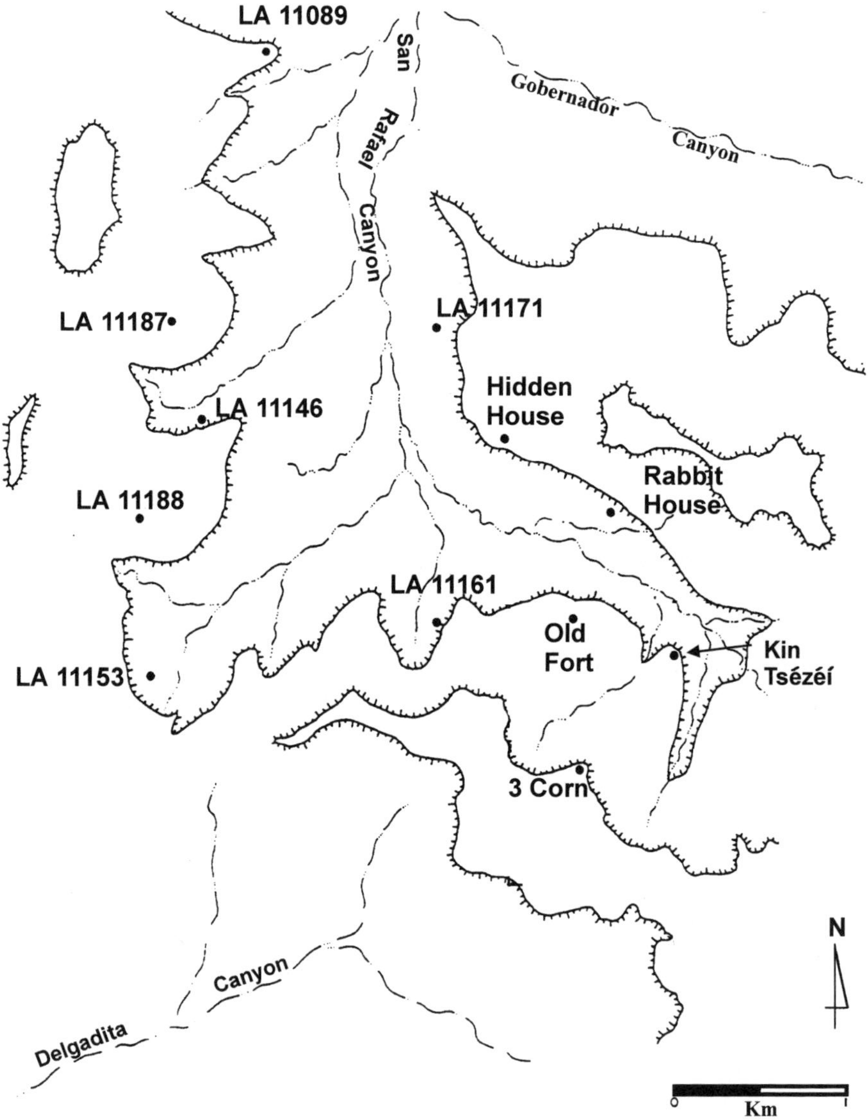

Figure 39. The San Rafael Canyon Dinétah Navajo communities

and becomes the La Jara drainage, a tributary to the San Juan. It is one of the few corridors that trends east–west rather than north–south. Two contemporaneous sites, Cabresto Mesa Tower and Pueblito East, sit on the north and south sides of the canyon, respectively (Figure 38). Cabresto Mesa Tower was built in 1713–1714 and Pueblito East may have been constructed in 1713; the single 1713r cutting date from Pueblito East is somewhat tenuous evidence for construction. Both sites are intervisible. Cabresto Mesa Tower sits high above the canyon and Pueblito East is located on the first bench on the south side of the drainage.

As discussed in more detail below, I believe that these two sites were part of a long-distance community designed to monitor this entrance to the Dinétah. Roque de Madrid's entrada entered the area via Vaqueros Canyon in 1705 (Hendricks and Wilson 1996), and it is probable that several later entradas used the same route. These sites may have been located to provide advance warning of Spanish troop movements that threatened the Navajo homeland.

The Blanco Canyon Community. Another of the earliest pueblito communities in the Dinétah was located in the west on Blanco Mesa. Three sites, Cottonwood Divide, Star Rock Refuge, and McKean Pueblito, comprise this far-flung association (Figure 40). The sites all date to the first and second decades of the eighteenth century, are in similar topographic settings, and are architecturally similar. In addition, McKean and Cottonwood Divide are intervisible.

Geographically, all three sites are located on, or adjacent to, Blanco Mesa; topographically, all are situated on high, isolated prominences that are extremely difficult to access. Architecturally, McKean Pueblito is the only site that contains intact masonry rooms, but it also contains somewhat informal "rooms" that were never roofed. Both Cottonwood Divide and Star Rock Refuge consist of forked-pole hogans, isolated wall segments, and covered crevices.

McKean Pueblito was built in 1708 and repaired in 1713. The tree-ring data from the other sites, however, are somewhat equivocal. The only date from Star Rock Refuge is a 1708vv noncutting date; Cottonwood Divide also failed to yield any cutting dates, but there is a weak cluster of noncutting dates in 1712–1715. The similarities between these three sites suggest that all may have been occupied at the same time in the 1710s. If that is the case, they might have formed a community on Blanco Mesa.

The Palluche Canyon Community. The Palluche Canyon community includes four, and possibly more, pueblitos (Figure 40). Twine House, 42 Pueblito, and possibly Foothold Pueblito were simultaneously occupied on the valley floor, and The Overlook Site and Compressor Station were

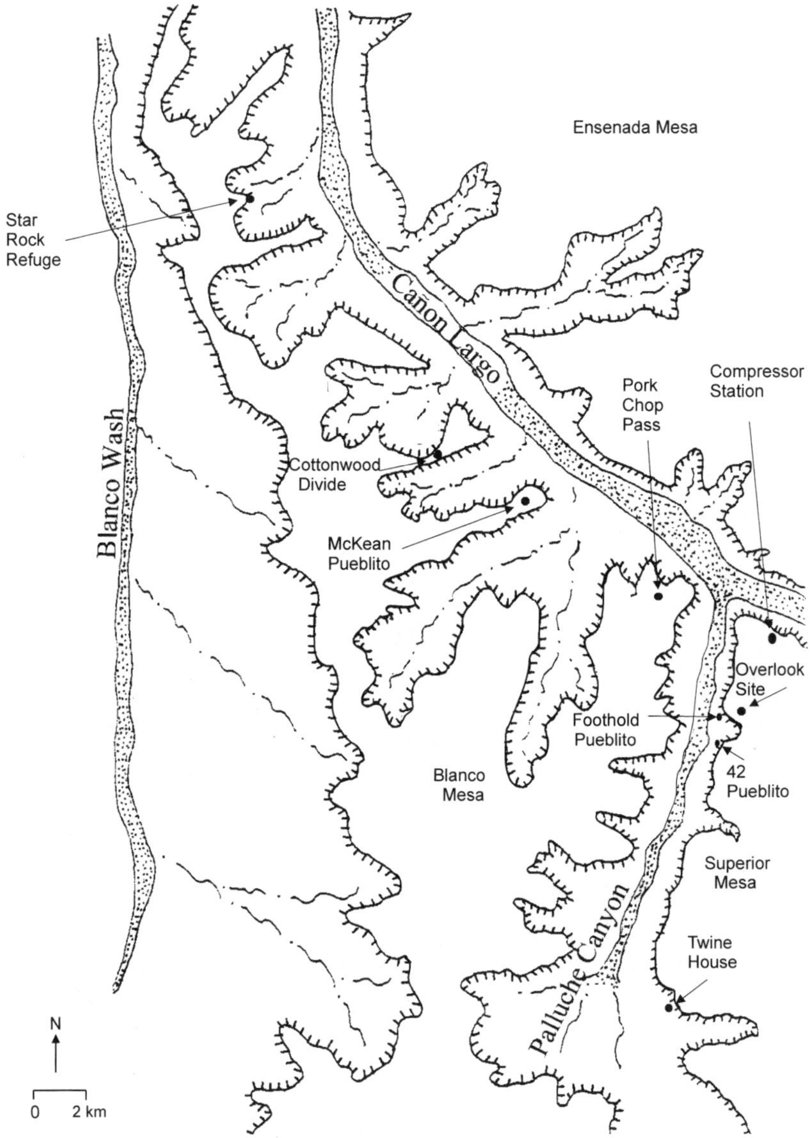

Figure 40. The Blanco Mesa and Palluche Canyon communities

built on the rim of Superior Mesa. Undoubtedly, there are numerous hogans and other structures that should be included in the community, but few have been identified and none has been dated. Comprehensive surveys (cf. Marshall 1991) in the area have yet to be conducted, but are planned for the floor of Palluche Canyon.

The tree-ring data indicate that there were two different occupations separated by abandonment of the pueblitos, if not of the entire canyon. Twine House, The Overlook Site, and Compressor Station were all built in 1727–1728; Foothold Pueblito and 42 Pueblito may date to the same time period, but the lack of cutting-date clustering prevents me from positively associating these pueblitos with the early Palluche Canyon community. The construction of Hooded Fireplace in 1722 and the 1727vv date from Split Rock Ruin may signify that the community extended southward and eastward across the Largo drainage. The configuration of the late community, occupied during the early 1740s, is somewhat different from earlier use in the area. Twine House, The Overlook Site, 42 Pueblito, and Foothold Pueblito all exhibit some evidence of use during the early 1740s, but Compressor Station appears to have been abandoned permanently in the late 1720s; the 1736–1737 dates from Largo School may represent a southward extension of this community toward the Largo.

The tree-ring data suggest that the Palluche Canyon community was initiated in the 1720s, depopulated in the 1730s, reoccupied in the 1740s, and abandoned by the early 1750s (Ababneh et al. 2001; Towner 1997). It is interesting to speculate on the relationship between the early Palluche Canyon community of the 1720s and the Blanco Canyon community of the late 1710s. Could the same group have founded both communities? Such a question can never be answered defninitively. However, Navajo oral traditions as described by Van Valkenburgh (1941), indicate that Palluche Canyon is the place of origin of the *Tsejikinii* and other clans (Copeland, personal communiction). Therefore, the entire area may have been used by the same extended kin group (clan?) for 30 to 40 years; Sesler, Hovezak, and Wilshusen (2000) suggest a similar land tenure system for the Frances Canyon area.

The Crow Canyon Community. The Crow Canyon drainage, a tributary to the Largo, contains one of the densest concentrations of pueblitos in the Dinétah. Seven dated sites are located in the area, but none contains strong clusters of cutting dates. Hadlock's Crow Canyon (1661vv) and The Citadel (1688++B) are poorly dated and only associated with the community on spatial criteria. Three sites, Boulder Fortress (1727–1728), Crow Canyon Pueblito (1723vv), and Ridgetop House (1720–1721) appear to be roughly contemporaneous, but none are intervisible. Finally, two rockshelter sites, Shaft House (1712+B) and Shándin Bíikin

(1705GB) are spatially associated with the sites, but may have been abandoned prior to 1720. If Shaft House was truly a refuge and not a habitation site, it may have been used sporadically by people who actually lived at Ridgetop House and/or Crow Canyon Pueblito.

The Frances Mesa/Morris 1 Community. Extensive and intensive survey projects on Frances Mesa (Sesler, Hovezak, and Wilshusen 2000) and in the Romine Canyon area (Dykeman and Wharton 1996) have documented substantial forked-pole hogan communities in both areas. Importantly, however, the survey areas abut each other and therefore may represent use by the same group(s) of Dinétah Navajos.

Only three dated pueblitos, Frances Canyon Ruin, Romine Canyon Pueblito, and Morris 1 Pueblito are present in the area, although undated pueblitos (Crevice Ruin, Romine Ranch Ruin, and LA 11091) are spatially associated with the community (Figure 38), and Quarter-Section Pueblito may be associated as well. As in other posited communities, this group exhibits a bimodal tree-ring date distribution. Beginning in the 1710s and continuing into the 1720s, Frances Canyon Ruin and Romine Canyon Pueblito were built and occupied. There is no tree-ring evidence that either site was used in the late 1720s or 1730s. Frances Canyon Ruin was apparently reoccupied in the 1740s, and Morris 1 Pueblito was built in 1748–1749. Romine Canyon Pueblito was apparently not occupied after the late 1720s. It is clear from the areal surveys, however, that Dinétah Navajos continued to use the landscape, even if there is no tree-ring evidence of pueblito construction.

The Central Cereza Community. What may be termed the "central Cereza" community (Figure 38) consists of at least two, and possibly five sites, and appears to be one of the later communities in the Dinétah. Hill Road Ruin and Gomez Canyon Pueblito were built in the 1730s and show evidence of remodeling and repair into the 1740s. If Room 4 at Gomez Canyon dates to the late 1740s, the site was occupied during the probable occupation of Adolpho Canyon Ruin which would make that site part of the same community. The 1749 dates from Gould Pass Pueblito suggest that it was part of the same late community, but its construction in 1751 may also indicate it was built after Gomez Canyon was abandoned. Finally, the 1736+vv noncutting date from Adams Canyon Pueblito suggests that it may have been part of the same community, but a single noncutting date is tenuous evidence, at best, to support this hypothesis.

The Southern Largo Community. There also may have been a more widespread community in the southern Largo area (Figure 38). Hummingbird House, Deer House, and Tower of the Standing God, all contain some evidence of use in the early and mid-1720s, and the two former sites also were reoccupied later in the 1740s. These three sites are widely separated, however, and are not intervisible.

Long-distance Interaction

On a much larger spatial scale, it is possible that a long-distance system of interaction exisited in the Dinétah, particularly in the early 1700s (Jacobson et al. 1992). In addition to Cabresto Mesa Tower and Pueblito East at the mouth of Vaqueros Canyon, Three Corn Ruin, Frances Canyon Ruin, and the sites on Blanco Mesa were probably occupied in the early 1710s. All of these sites are intervisible *as a network*, although not individually. LA 12929, an undated site in the south, and Pointed Butte, a poorly dated site south of Superior Mesa, also may have been part of this extended network.

There are two important implications of this network. First, a variety of communication methods would have been necessary between individual sites (Lange 2001). Second, if such a long-distance network was operational, the Dinétah Navajo social organization of the early 1700s must have been more complex than has been assumed in the past. Whether or not this organization was truly "tribal" in nature remains problematic. The evidence suggests that this system was not operational during the later (1730s–1750s) occupation of Dinétah, which raises the question, "why not?" Was there some change in Navajo social organization that facilitated a more local defensive orientation, or was the change simply a result of responding to different threats? Such questions remain unanswered.

CHAPTER 5

Climate, the Navajo Occupation, and Abandonment of the Dinétah

Abandonment conjures up images of catastrophe, mass migration, and environmental crisis (Cameron 1995:1).

The tree-ring dates, historical records, and other data indicate that the Navajo ceased to inhabit the Largo-Gobernador area after about A.D. 1760. Although the area may have been used for ceremonies—something that still occurs today—and it may have been sparsely used during the Bosque Redondo period (Towner and Johnson 1998), there are no habitation sites that postdate A.D. 1755. Because of a variety of factors, primarily the nature of the tree-ring and historical data that have been analyzed, the Navajo abandonment of the Dinétah has indeed evoked images of a mass migration precipitated by both social and natural environmental crises. This chapter examines several concepts related to the abandonment including the climate before, during, and after the pueblito site occupations, the changing nature of Navajo settlement during the pueblito occupations, and the relationship between climate, the pueblito occupations, Ute and Comanche raiding, and areal depopulation. It also discusses the timing and destinations of possible emigrants.

CLIMATE

Climate has long been a topic of interest to Southwest archaeologists. Although there are some notable exceptions (Brugge 1986; Kemrer 1974; Oswald 1993), climatic factors usually have not been invoked as explanatory tools by students of early Navajo archaeology. There are three principal reasons for this omission. First, most published paleoclimatic data, usually tree-ring reconstructed precipitation or temperature data (Dean and Robinson 1977), have presented decadal summaries of departures from mean conditions for relatively large areas of the northern Southwest. Similarly, the low-frequency processes (LFP) that have been of most concern to archaeologists studying the Anasazi (Dean 1988) have generally been viewed as covering too long a time span to have affected Navajo settlement patterns. Although useful (cf. Powers and Johnson 1987:3), the decadal summaries have not had the temporal or spatial resolution necessary to address the questions of concern to archaeologists studying early Navajo development. Second, Navajo subsistence in the historic period was not exclusively oriented toward maize agriculture and, therefore, most researchers have assumed the Navajo were less affected by environmental fluctuations than their Puebloan neighbors. Most archaeologists and historians have viewed the Navajo transition to pastoralism as an adaptation to the social, not natural, environment (Bailey and Bailey 1986). Finally, because the Navajo sites of interest were occupied during the historic period, historical accounts and anecdotal references have been used by most scholars in discussions of environmental oscillations and their effects on the Navajo (Reeve 1959). These data, while useful, have not been used to generate inferences concerning the patterns and trends of climatic fluctuations in relation to Navajo settlement in the Dinétah.

A fine-grained reconstruction of precipitation in the Gobernador area has been developed recently by Gary Funkhouser of the Laboratory of Tree-Ring Research. Using live-tree cores from Pueblito Canyon, archaeological samples, and precipitation data from the Santa Fe divisional weather database, he generated an annual reconstruction of precipitation for the Gobernador area for the period A.D. 624–1975. Several caveats must be stated prior to a discussion of climate in the area. First, the tree-ring reconstruction is for precipitation only; temperature is undoubtedly an important component of climate (Salzer 2000), but was not used in the reconstruction. Palmer Drought Severity Indices (PDSI), which incorporate both precipitation and temperature (Palmer 1965), may prove useful in the future, but were not used, primarily because temperature has a strong inverse relationship to precipitation in this area. Tree-ring reconstructions of precipitation have limits that should be acknowledged (D'Arrigo and Jacoby 1991). Both high and low values may be underestimated, high values because moisture ceased to limit tree growth during extremely wet years, and low values because ring widths cannot fall below zero no matter how little rain falls in dry years (Fritts 1991). The precipitation reconstruction is presented as annual values in inches for the "tree-year," previous August through current July. Because the Dinétah receives most of its precipitation as winter snow, the annual reconstruction, like the instrumental annual values upon which it is calibrated, does not separate seasonal components well. Certainly summer precipitation, particularly convection thunderstorms in July and August, were important for some types of agriculture. Winter precipitation, however, provides moisture during the early months of the growing season, a factor that may be critical for agricultural and wild plant foods, as well as for animal forage. Thus, the annual precipitation reconstruction is considered adequate for comparing climate with human events measured by the archaeological tree-ring data. Finally, in this discussion as in many others, I assume that years with high precipitation values, those more than one standard deviation above the mean, were "good" years, and low precipitation years, those more than one standard deviation below the mean, were "bad" years. Although such may not always be the case (cf. Graybill 1989), given the presumed subsistence base of the early Navajo (Dykeman 1998), I believe it justified.

Long-Term Climatic Trends in the Dinétah

To better illustrate the trends, Figure 41 shows the reconstructed precipitation for the period A.D. 1400–1775 both as yearly values and as a five-year running mean. It is during this time that most scholars assume the Navajo moved into and occupied the area. The long-term mean for this period is 10.90 inches of precipitation per year with a standard deviation of 1.936; the highest reconstructed value is 16.858 inches in 1485, and the lowest is 5.318 inches in 1748. Obviously, precipitation has varied tremendously in the area over the 500-year-long period of interest. The Navajo occupants of the area, however, undoubtedly adapted to such variability in a number of ways.

Several trends may be important in relation to the Navajo entry into the area. The 1300s, particularly the first three decades of the century, were a time of relatively low temporal variability and generally good conditions. Precipitation exceeded one standard deviation above the mean in twelve years during the 1300s, including seven years in the first half of the century. In ten years, seven of which were in the last half of the century,

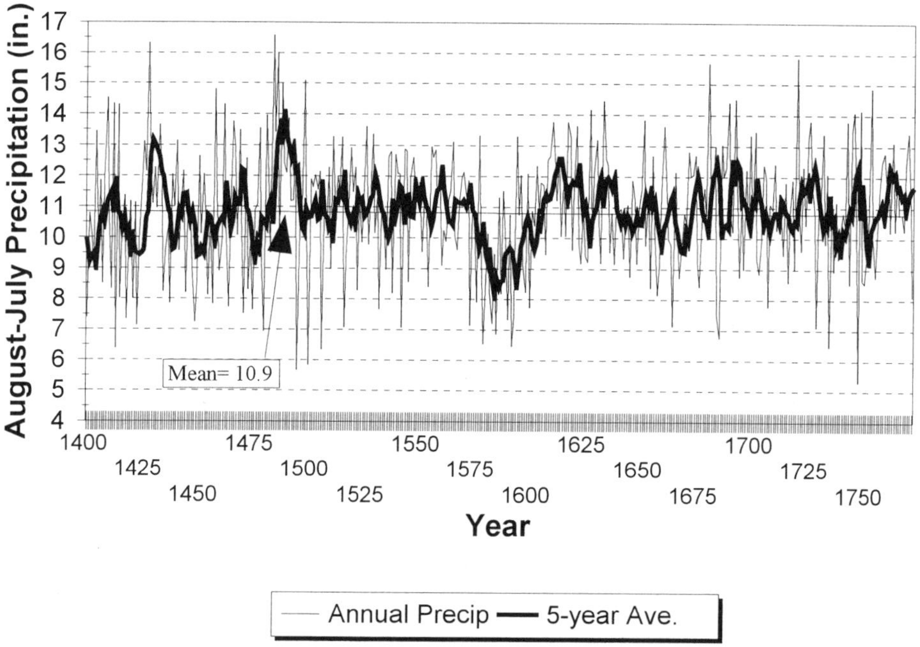

Figure 41. Reconstructed Gobernador precipitation, A.D. 1300–1800

precipitation was at least one standard deviation below the mean. Thus, if the Navajo moved into the area in the early fourteenth century as proposed by some researchers (Reed and Horn 1990), they met with generally favorable and somewhat stable conditions. Even if they entered the area in the late 1300s, conditions were generally stable, although not as favorable as early in the century.

The 1400s were a time of high temporal variability and sometimes extreme departures from "normal" conditions. Both "good" and "poor" years occurred 23 times during the century. Good years occurred more often (n=14) in the later half of the century, including three "very good" years (>2 standard deviations from the mean) in the 1480s. Poor years happened more frequently (n=13) in the first half of the 1400s, although two extremely poor years, 1480 and 1495, were in the later decades. If the Navajo penetrated the area in the 1400s (Hodge 1895), therefore, they entered during a time of unstable conditions when precipitation varied substantially from year to year and decade to decade.

The 1500s were a time of low variability, but also a time of generally low precipitation as well. There were more than twice as many bad years (n=23) as good years (n=10) during the century. There were no very good years, and the majority of the good years (n=8) were prior to 1575. Bad years, on the other hand, included five very bad years, and occurred more often in the later part of the century. There was a significant downturn in precipitation beginning in the 1570s that continued until the end of the century. In the last quarter of the 1500s, only 1577 and 1594 were good years, whereas there were 11 years with below average precipitation, three of which were "very bad" years. This severe drought in the late sixteenth century has been well documented (Parks et al., n.d.) and may have affected other populations in the Southwest. Thus, if the Navajo entered the Largo-Gobernador area during the 1500s, particularly in the last quarter of the century, they met with stable, but very poor conditions.

Climatic conditions improved markedly in the 1600s. Temporal variability increased from the previous century, and precipitation generally increased substantially. Good years (n=15) outnumbered bad (n=9), and only two years were extreme—1680 was a very high precipitation year and 1685 a very low precipitation year. After the drought of the late 1500s and a poor year in 1600, conditions ameliorated for the next 45 years; between 1600 and 1645, there were seven good years and not a single bad year. Although the last half of the century was somewhat more variable and conditions were not quite as favorable, the 1600s were a marked improvement over the 1500s. If the Navajo began to occupy the Dinétah during the 1600s, they met with generally favorable precipitation and relatively predictable conditions. It may be significant that some of the well-dated forked-pole hogans near the San Juan River described by Sesler and Hovezak (1996) were built in the first few decades of the 1600s. A Navajo expansion southward into unoccupied territory at that time would have been facilitated by rising water tables (Dean 1988) and more stable and favorable conditions. In the 1620s, which were generally a "good" decade, the Navajo were described as having "vast corn fields" (Lummis 1900) and trading wild animal products at Santa Clara (Worcester 1947).

Climate during the Navajo Occupation of Pueblitos

Figure 42 shows the yearly precipitation for the tree-ring-dated period of Navajo occupation, A.D. 1610–1760. This period is important because it is the time when confirmed tree-ring-dated Navajo sites have been documented and, later, when the pueblitos were built and occupied. Figure 42

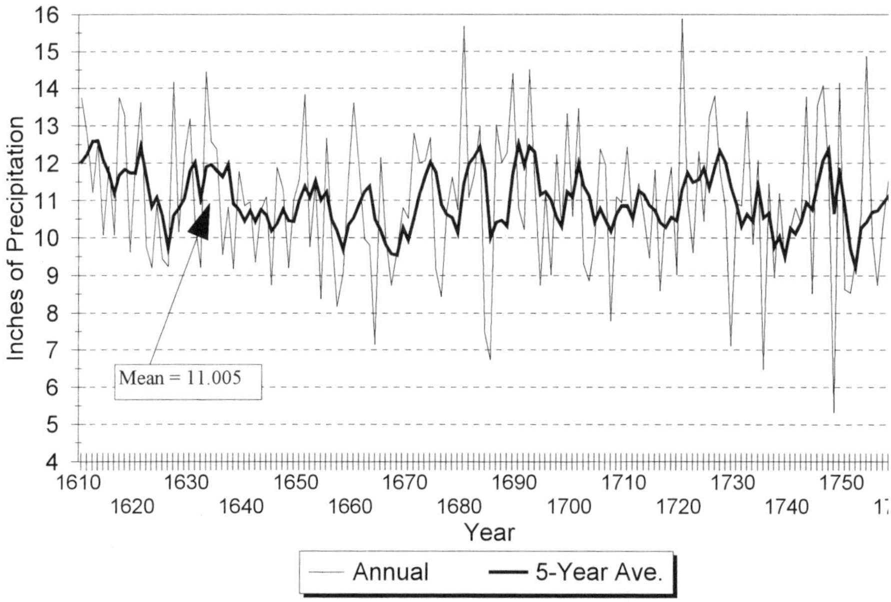

Figure 42. Reconstructed Gobernador precipitation, A.D. 1610–1760

also shows a three-year moving average of precipitation superimposed over the yearly values. Precipitation fluctuated from a low of 5.318 inches to a high of 15.878 inches of moisture; the mean annual precipitation for the period is 11.005 inches, and the standard deviation is 1.912 inches.

Table 19 lists the "good" (wet) and "poor" (dry) years for the 1610–1760 period. As can be seen, there were more good years (n=25) than bad years (n=22), but more extremely bad years (n=4) than extremely good years (n=3). After the Buffalo Mask site was constructed in 1680, the frequency of good and bad years was the same until 1760 (n=16); there were three very good years (1680, 1720, and 1754), and four very bad years (1685, 1729, 1735, and 1748). The variability remains approximately the same if only the major period of pueblito construction (1710–1755) is considered; good and bad years occurred with the same frequency (n=9), but there were three very bad years and only one very good year. Overall, the reconstructed precipitation data indicate that during the pueblito occupation temporal climatic variability was relatively high; although there were many bad, and some very bad years, there were also many good, and a few very good years.

There are several ways to assess the impact of climate on the Navajo

Table 19. List of Good and Poor Precipitation Years, A.D. 1610–1760

Good Years		Poor years	
Year	Inches	Year	Inches
1610	13.764	1645	8.740
1617	13.764	1654	8.381
1618	13.290	1657	8.165
1621	13.643	1658	9.035
1627	14.170	1664	7.142 **
1630	13.195	1667	8.724
1633	14.449	1676	8.439
1651	13.843	1684	7.453
1660	13.616	1685	6.736 **
1680	15.683 **	1694	8.740
1683	13.000	1703	8.850
1686	13.036	1707	7.785
1689	14.402	1716	8.581
1692	14.507	1719	9.019
1699	13.337	1729	7.111 **
1701	13.469	1735	6.473 **
1720	15.878 **	1737	8.935
1725	13.232	1744	8.518
1726	13.806	1748	5.318 **
1732	13.390	1750	8.624
1743	13.785	1751	8.534
1745	13.548	1756	8.740
1746	14.086		
1749	14.154		
1754	14.866 **		

Mean = 11.00534 inches
** = 2 Standard Deviations from Mean

occupation. Brugge (1986:157–60) correlated tree-ring indices with dated historical events for the Navajo occupation of the Chaco Canyon area. Unfortunately, of the 50 historical events he listed, only six were during the time period when pueblitos were occupied. In addition, the vagueness of some "events," such as "continued war" or "land disputes," makes precise dating problematic. Thus, it is difficult to find a strong correlation between climate and historical events, and Brugge (1986:160) suggests that "environmental change may lead to change in interethnic relations, but that the direction and nature of change can be strongly conditioned by cultural factors."

Given the infrequent and imprecise nature of historical references to the pueblito occupations (Hendricks and Wilson 1996), it is much more informative to compare the archaeological data and dendroclimatic reconstruction. Using only cutting and near-cutting dates from the pueblito

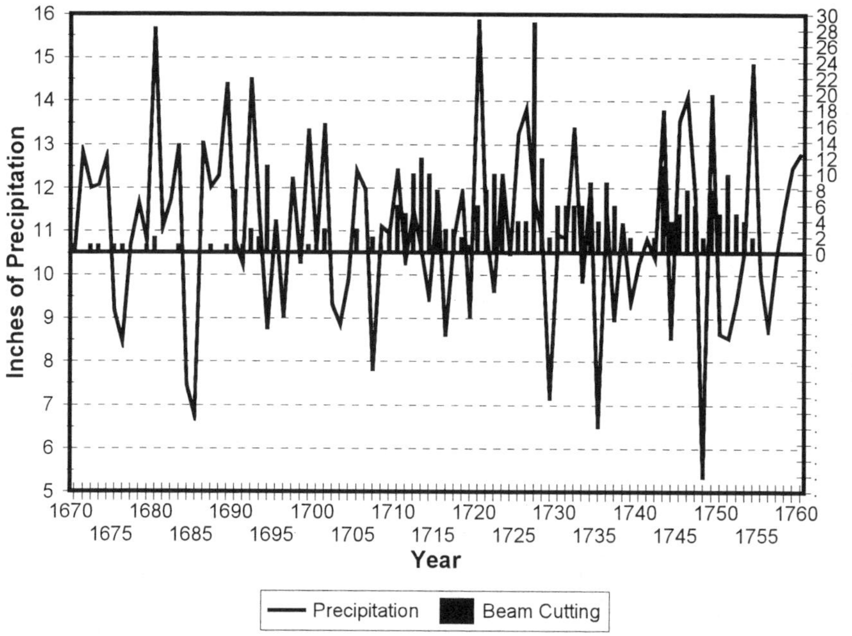

Figure 43. Temporal distribution of tree cutting and precipitation, A.D. 1670–1760

sites, Figure 43 shows the temporal relationship between tree cutting and reconstructed precipitation. There were 66 years between 1670 and 1760 when tree cutting occurred. Not surprisingly, most trees were cut in "average" years; this observation is simply a reflection of the fact that there are more average years than either good or bad years in the time period. Precipitation declined from the previous year 35 times, and 175 beams were cut during those years; precipitation increased in 31 years, when 140 beams were cut. These data suggest that tree harvesting may have been tied to declining precipitation; however, because most are "average" years, declines or increases from the previous year are often minor fluctuations or may be above-average high values. Thus, the direction of precipitation change may be unimportant in most years when "average" conditions prevailed.

Trends in tree harvesting during extreme years, however, are more informative. Overall, 49 beams were cut in the twelve bad years between 1670 and 1760 when trees were cut (no trees were cut in four bad years); during the same time, 64 beams were cut in the fifteen good years (no trees were cut in two good years). Prior to 1690, the sample of trees cut is too small to make any generalizations. After 1690, however, tree cutting

Table 20. Tree-Cutting Activities in Poor and Good years

Poor Years			Good Years		
Year	Tree Cutting (n=)	Tree Cutting Change	Year	Tree Cutting (n=)	Tree Cutting Change
1694	11	+9	1692	3	+2
1696	0	NC	1699	1	+1
1703	0	NC	1701	3	+2
1707	1	+1	**1720	6	+5
1716	3	-5	1725	4	+3
1719	1	-1	1726	4	NC
**1729	2	-10	1732	6	NC
**1735	4	-5	1743	11	+9
1737	5	-3	1745	4	NC
1744	4	-7	1746	8	+4
**1748	2	-4	1749	8	+6
1750	5	-3	1754	2	-2
1751	10	+5			
1756	0	NC			

Poor Years (n=14)
Tree Cutting Increase (n=3)
Tree Cutting Decrease (n=8)
Tree Cutting Unchanged (n=3)
**=2 Standard Deviations from Mean

Good Years (n=12)
Tree Cutting Increase (n=8)
Tree Cutting Decrease (n=1)
Tree Cutting Unchanged (n=3)

is an activity that seems to correlate positively with above average precipitation and negatively with below average precipitation (Table 20). Between 1690 and 1760, there were 14 bad years; tree-cutting activities declined in eight of those years, increased in three years, and remained unchanged from the previous year three times. Interestingly, the three unchanged years were times when no trees were cut in either the bad year or the previous year. During the same time period, there were 12 good years; tree-cutting activities increased in eight of those years, remained unchanged three times, and decreased only once. Notably, the only good year with a decrease was 1754, the last year there was tree harvesting near any of the pueblito sites. These data clearly indicate that tree harvesting was conducted in average and above-average years; it was done in below-average years, but not as frequently.

Another way to assess any relationship between the pueblito site occupations and climate is by comparing construction events and precipitation (Figure 44). In this comparison, the construction events are identified to the year based on the data presented in Chapter 3; because the Buffalo Mask site is included, precipitation data are extended back to 1680.

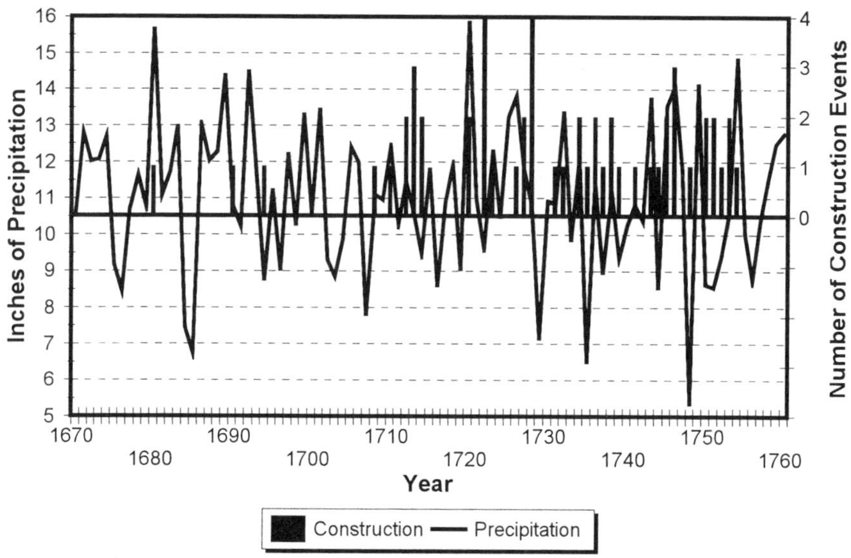

Figure 44. Temporal distribution of construction events and precipitation, A.D. 1670–1760

Overall, pueblito construction occurred in 32 of the years between 1680 and 1760; 17 of those years were when precipitation increased from the previous year, and 15 were when it decreased. In this time period there were 14 bad years, including four very bad years, and 13 good years, including only two very good years (Table 21). In most of the bad years (n=9) pueblito construction remained unchanged from the previous year; it decreased twice (1729 and 1737) and increased three times (1694, 1748, 1750). In the good years, construction remained unchanged six times, declined twice (1749, 1754), and increased in five years (1680, 1720, 1726, 1743, 1746). Two of the three bad years when pueblito construction increased may be special cases. The first, 1694, is the construction of Tapacito Ruin, which must be considered separately from other pueblitos (Towner and Dean 1992; Towner 1996). The second, 1748, is defined by the weak cluster of near-cutting dates from Morris Site 1. These dates, 1748+G and 1748+GB, are both near-cutting dates; it is highly probable that these trees were actually cut in 1749 (a good year), and that the possible missing ring indicated by the "+" is 1748. The only other construction episode in a bad year, 1750, was part of the major occupation of Old Fort Ruin. Finally, the only two good years when pueblito construction declined include 1749—a year when no pueblitos were constructed, and 1754—the final year of pueblito construction. It

Table 21. Pueblito Construction Trends in Poor and Good Years

Poor Years			Good Years		
Year	Pueblito Construction (n=)	Change	Year	Pueblito Construction (n=)	Change
1684	0	NC	1680	1	+1
1685	0	NC	1692	0	NC
1694	1	+1	1699	0	NC
1696	0	NC	1701	0	NC
1703	0	NC	**1720	2	+2
1707	0	NC	1725	0	NC
1716	0	NC	1726	1	+1
1719	0	NC	1732	1	NC
**1729	0	-3	1743	1	+1
**1735	1	NC	1745	1	NC
1737	1	-1	1746	3	+2
1744	1	NC	1749	0	-1
**1748	1	+1	1754	1	-1
1750	2	+2			
1751	2	NC			
1756	0	NC			

Bad Years (n=16)
Construction Increase (n=3)
Construction Decrease (n=2)
Construction Unchanged (n=11)
**=2 Standard Deviations from Mean

Good Years (n=13)
Construction Increase (n=5)
Construction Decrease (n=2)
Construction Unchanged (n=6)

must be remembered, however, that the Morris 1 pueblito may have been built in 1749. These data, like the tree-harvesting data, suggest that pueblitos were not built in bad years.

In a sense, pueblito construction in good years, especially if pueblitos were built for defensive purposes, seems somewhat counterintuitive. Why should defensive needs peak in times of above average precipitation? If the enemy was the Ute or Comanche, as suggested by several researchers (Carlson 1965; Towner 1992, 1996; Towner, Rogers, and Copeland 2001), the attackers and defenders of the pueblitos may have been under similar environmental stresses. Yet the tree-ring data indicate construction in years that would have been relatively "stress-free." In this context, it is important to identify the goals of those who were attacking pueblitos. Unlike the Spaniards whose goals were to "punish" the Navajo for previous raids, recapture livestock, and "rescue hostages" (Hendricks and Wilson 1996), the Ute were primarily interested in capturing property, stock, and especially, slaves (Jacobson et al. 1992). Indeed the emerging slave market in Mexico may have been one of the primary incentives for much of the raiding by all parties across the Southwest (Towner, Rogers, and Copeland 2001).

I suggest that Ute, not Navajo behavior, explains why most pueblitos were built in good years. Probably the most important factor is that the Ute, or other non-Spanish people in the area, were raiding primarily for slaves, although livestock and other goods may have also been stolen (Jacobson et al. 1992). In good years, the slave raiders may have had an ample supply of food, so that if the raid was unsuccessful, their subsistence base was not adversely impacted by scheduling conflicts or incurred costs of the raid. If it was successful, on the other hand, the raiders also had the extra food on hand necessary to feed their captives. In bad years, however, the raiders may not have had the necessary "buffer," in terms of stored subsistence items, to attempt such an endeavor. Simply put, in good years, the raiders could profit by the venture but—other than bodily harm—were actually risking very little. In bad years, however, the risks of failure outweighed the possible rewards of success.

Although raiding behavior has not been studied extensively, there are some similar examples of when raiding occurs. The Iban of the Philippines have been traders and raiders, as well as part-time swidden agriculturalists, for much of their history (Gibson 1990). In the nineteenth century, raiding was conducted primarily to acquire slaves for trade with other Southeast Asian societies and to acquire status for young, unmarried males. Iban agriculture required the labor of these males for only two or three months of the year, leaving the remainder of the year "to free young men for the activities of trading and raiding" (Freeman 1970:222).

When the Apache raided for livestock, on the other hand, they were apparently responding to a shortage of meat, although other items are not mentioned; they raided "whenever it was apparent that the meat supply . . . was running low" (Goodwin 1942:16); there is no indication, however, that slave raids were conducted at the same time. In terms of Navajo raiding, Hester (1962a:76) lists 34 different references to Navajo attacks on either the Spaniards or the Pueblos; five entries include multiple years, and six entries are multiple events that occurred in the same year. Of the 22 individual years noted for Navajo raids, most were "average" years; only three raids occurred in bad years, and one occurred in a good year. The only good year of a Navajo raid was 1793, when there were hostilities with the Gila Apache. The three bad years include 1694, 1780, and 1788; the last was a combined Navajo-Spanish expedition against the

Gila Apache, and the 1780 raids indicate conflict with both the Hopi and Acoma. Thus, 1694 again stands out as a unique year in the Navajo occupation of the Dinétah.

These data suggest that most pueblitos were built in response to Ute and/or Comanche raiding. Furthermore, the correlation between climate and pueblito construction suggests that those raids occurred primarily in above-average years. I suggest that raiding for slaves occurred primarily in good years because the raiders had a surplus in case the raid failed. Thus, the old adage that "desperate times require desperate measures," does not appear to have applied to slave raiding in eighteenth-century New Mexico, at least as far as the Native American groups were concerned. It may have applied to raiding for food, but the Navajo subsistence base was such that agricultural products or the few available sheep would not have been vulnerable to theft by "galloping into an enemy camp and grabbing the goods and women before an adequate defense could be organized" (Jacobson et al. 1992:110).

Questions remain, however, concerning why the Ute/Comanche began raiding in the early 1700s. Most historians trace the rise of Ute raiding to the availability of both horses and guns in the late seventeenth and early eighteenth centuries (Secoy 1992; John 1975). Although some horses were probably available prior to 1700, the Ute apparently were not mounted. In the early 1700s, however, Ute horse stealing became more common (Thomas 1935). The displacement of many Spanish horses during the Revolt and Reconquest may have also contributed to the Ute acquisition of the horse. The Navajo too were able to acquire horses, but as at least part-time horticulturalists, they had less need for them; Navajo horses may have been one of the goals of Ute raids. It was against official Spanish policy, however, to sell or trade guns to Indians, and neither the Puebloans nor Navajo were heavily armed in the early 1700s. By contrast, the Plains Apache, and probably the Ute as well, were apparently being supplied with French guns via a trade route up the Arkansas River (John 1975). When Juan de Ulibarri met the Cuartelejo Apache in 1706, they were anticipating an attack from the Ute and Comanche; the Cuartelejos asked for a gun "because the French were now giving guns to their enemies" (John 1975:229). Thus, it appears that in the early 1700s, the Ute possessed horses that increased their mobility and guns that increased their firepower. With a ready market for Navajo slaves and a need for additional horses and other goods, all the ingredients were in place for a dramatic increase in Ute raids aimed at Navajo rancherías. The most effective defense against such raids was pueblitos, both in terms of their locations and their construction. If the targets of the raids were hidden inside pueblitos located on boulders and near mesa edges, the attackers were forced to dismount in order to capture either slaves or goods—a strategy which eliminated the aspect of speed in the raid and left the attackers vulnerable to counterattacks.

Climate and the Abandonment of Dinétah

Several possible reasons for the Navajo abandonment of the Dinétah have been forwarded over the years. These "pushes" (Anthony 1990) have never been quantitatively defined but include drought and pressure from Ute and Comanche raiding. Drought has been assigned a significant role in the abandonment (Marshall 1995), and the drought of 1748 in particular (Reeve 1958) has been suggested as a cause. Ute and Comanche raiding has also been suggested as the reason for the abandonment (Schroeder 1965:59), as has a combination of both drought and raiding (Carlson 1965; Brugge 1972b). Finally, the shift to a pastoral economy and the need for better grazing and herding areas, precipitated by drought and Ute raiding, has also been proposed as both a cause and a result of the Dinétah abandonment (Haskell 1975).

The correlation between tree cutting and precipitation, and the correlation between pueblito construction and precipitation, suggest that drought had little influence on the abandonment of the area. Certainly, 1748 was a very dry year; indeed, in the Gobernador area precipitation reconstruction it is the driest year in the entire 1348-year chronology. However, two of the preceding three years, 1745 and 1746, were above-average years, and the succeeding year, 1749, was the wettest in almost thirty years. Thus, the "drought" of 1748, which supposedly made the Navajos "amenable to the missionaries' suggestion that they move south to the Cebolleta region" (Reeve 1959:20; Hester 1962a:77) was a single-year event. Other dry years, such as 1729 and 1735, have not been given such importance by those inferring climate as an abandonment push.

In addition to the climatic data, the archaeological data indicate that pueblitos were constructed after the supposed drought. At least seven construction episodes, and possibly several more, at as many as eight sites were completed in the years following 1748. Pueblito construction actually expanded into new areas, such as north of the San Juan River, during this time. Finally, those that advocate that a precipitation deficit was a critical factor in the abandonment of the Dinétah must explain why the latest construction episodes occurred in 1754, the wettest year since 1720 and second wettest since the first pueblito was constructed in 1680. It is apparent from these data that climate was not one of the "pushes" behind the Navajo exodus from their homeland. Instead, Ute and/or Comanche raiding (Hester 1962a; Reeve 1958) and possibly free-lance Spanish slave raiding (Bailey 1964) appear to have been the primary push factors behind the Navajo emigration.

Before discussing the destinations of Navajo emigrants and the possible "pulls" that attracted them to various areas, it is necessary to discuss the nature of the Navajo occupation in the late 1740s and early 1750s. As discussed in the previous chapter, seven sites contain evidence of occupation in the late 1740s; four of these sites, including three in San Rafael Canyon, also show evidence of occupation in the early 1750s. Only three newly constructed sites, Rabbit House, Gould Pass, and Truby's Tower, were occupied in the early 1750s. Towner and Johnson (1998) indicate that Rabbit House was part of the Old Fort community in San Rafael Canyon; Gould Pass may have been part of the central Cereza community; there is no evidence of a community in the south near Truby's Tower or in the upper Largo. Thus, the final abandonment of the area after 1755 may have involved very few people—only those living in San Rafael Canyon, on top of Ensenada Mesa, and near the confluence of Tapacito Creek and Cañon Largo. If there was ever a "large" Navajo population in the area, a hypothesis that the episodic nature of building construction does not support, emigration must have begun long before the 1740s and 1750s. If, as indicated by the climatic data, drought was not one of the "push factors" (Anthony 1990; Ahlstrom et al. 1995) in the Navajo emigration, I suggest that Ute/Comanche raiding was the primary push factor, and that it began having an impact in the 1720s, if not earlier.

Although no sites in the Dinétah heartland date between 1760 and Bosque Redondo (1860s), a very interesting and potentially significant group of sites northeast of Aztec, New Mexico, deserves mention. Fetterman (1996) reports a 1762v date from a roasting pit at LA 81172. Three Navajo Land Claim sites, N-SJ-LA-D, E and J have yielded noncutting dates ranging from 1530++vv to 1775vv; the earlier dates are obviously dead wood, but the later dates suggest a post-1760 Navajo use of the area, possibly centered on NLC site N-SJ-LA-J (Figure 45), known locally as the "Navajo Fort." Unfortunately, the only dates from this site are 1530++vv, 1576++vv, and 1627++vv. Only N-SJ-LA-E has yielded a

Figure 45. Pueblito site NLC-SJ-LA-J (The Navajo Fort)

cutting date, 1746v, but N-SJ-LA-D has yielded two noncutting dates (1771vv and 1775vv) that suggest an occupation in the last few decades of the eighteenth century. This is one area where additional dendrochronological, archaeological, and oral traditional data could contribute significant new information regarding the depopulation of the Dinétah.

If the Navajo had depopulated the Largo-Gobernador area by the 1760s primarily because of raiding, two questions remain. First, what were the destinations of the migrants, and second, what "pull" factors enticed them to new areas? Hester (1962a:84) indicates movement to the south and west, toward the Cebolleta Mountains and Canyon de Chelly. Although I agree with this general trend, the timing, number, and variety of destinations need to be revised.

South of the Largo-Gobernador area, there are three major concentrations of known Navajo archaeological sites, the Big Bead Mesa (Keur 1941), Cebolleta Mesa (Carroll 1979), and Chacra Mesa areas (Brugge 1986; Vivian 1960). Big Bead Mesa has been known for many years and was the type site of the Cabezon Phase (Hester 1962a). Big Bead Mesa itself contains tree-ring dates from the last half of the eighteenth century

Table 22. Non-Dinétah Navajo Sites in New Mexico with Pre-1800 Tree-ring Dates

Site Name or Number	Number of Dates	Earliest Cutting Date	Latest Date	Number of Date Clusters
NLC-S-ULC-CZ-A	20	1510	1834	0
NLC-S-AS-AC-T	1	1628	1628	0
NLC-S-ULC-CZ-C	9	1655	1764	0
NLC-S-ULC-EM-Z	9	1701	1850	0
NLC-S-AS-AM-B	6	1704	1800	0
NLC-S-ULC-UP-MM	4	1721	1762	0
SITE CM-18	3	1725	1745	0
NLC-E-C-UC-N	1	1725	1725	0
SITE CM-38	19	1738	1972	1
STAR LAKE	1	1739	1739	0
LA 5228	2	1747	1752	0
NLC-S-ULC-UP-LL	14	1755	1764	1
BIG BEAD MESA	33	1766	1856	1
NLC-E-RP-MP-W	5	1787	1787	0
NLC-E-RP-CM-K	3	1793	1863	0
NLC-S-ULC-ZR-DD	5	1793	1857	0
NLC-S-ULC-UP-JJ	3	1794	1802	0
NLC-S-AS-AC-BB	5	1796	1797	0
NLC-E-RP-CM-C	1	1796	1796	0

(Table 22). The site has often been considered in isolation; however, a recent analysis of site survey forms of the Anasazi Origins Project and other small projects in the area, lead me to conclude that at least 29 sites near Big Bead Mesa predate A.D. 1800 (Towner 1994). Unfortunately, the data are currently inadequate to refine the temporal association of most of these sites, and the survey coverage is spotty at best. I have no doubt that more, and probably earlier, Navajo sites are located in the area. LA 5228 is one site that certainly predates the abandonment of the Largo-Gobernador area.

A portion of the Cebolleta area near Mount Taylor was surveyed in the late 1970s (Carroll 1979). The survey encompassed approximately 8.45 square miles; 105 Navajo sites were recorded, including at least 299 hogans of various types. Carroll suggests that many of the sites postdate A.D. 1800, but Marshall's (1979) ceramic analysis indicates that at least 11 predate 1750 and two may predate A.D. 1700. Tree-ring data from these sites are limited and include no cutting dates.

The best dated post-Revolt pre-1750s Navajo occupation outside the Largo-Gobernador area is on Chacra Mesa. Through the work of both Vivian (1960) and Brugge (1986), it is obvious that several of the pueblito

Table 23. Navajo Sites in Arizona with Pre-1800 Tree-ring Dates

Site Name or Number	Number of Dates	Earliest Cutting Date	Latest Date	Number of Date Clusters
NLC-W-LLC-TS-E	1	1666	1666	0
ARIZONA D:7:4166	8	1691	1890	0
NLC-W-LLC-NJ-B	1	1703	1703	0
NLC-W-LLC-P-F	1	1710	1710	0
NLC-W-LLC-OP-T	1	1711	1711	0
ARIZONA D:11:361	11	1713	1941	0
ARIZONA D:7:276	6	1719	1777	0
NLC-W-LLC-MJ-SSS	9	1720	1882	0
NLC-W-LLC-J-J	1	1723	1723	0
ARIZONA D:7:4163	5	1726	1926	0
NLC-W-LLC-UO-O	2	1728	1817	0
NAZLINI CANYON	18	1751	1762	1
NLC-W-LLC-C-OO	29	1756	1830	2
KINNAZINDE	16	1759	1760	1
NLC-S-MLC-LP-O	8	1762	1764	1
NLC-W-LLC-C-F	5	1765	1809	0
NLC-W-LLC-SM-X	2	1766	1860	0
ARIZONA D:7:228	30	1768	1962	2
NLC-W-LLC-DO-G	2	1772	1788	0
NLC-W-LLC-MJ-DD	4	1772	1806	0
CC 351 (NPS)	45	1773	1773	0
ARIZONA D:7:504	4	1775	1865	0
ARIZONA D:12:213	7	1775	1891	2
NLC-S-ULC-UP-R	4	1777	1802	0
ARIZONA D:7:534	8	1782	1812	0
NLC-W-LLC-C-B	11	1793	1803	1
ARIZONA D:12:215	12	1793	1862	1
NLC-W-LLC-UO-Q	1	1795	1795	0
NLC-W-LLC-C-CC	2	1795	1828	0
NLC-W-LLC-MJ-III	1	1797	1797	0
NLC-S-ULC-UP-P	1	1798	1798	0
NLC-W-LLC-C-E	8	1798	1839	0

sites in the area were occupied before the Largo-Gobernador area was abandoned. The Shepherd site, variously recorded as NLC-E-C-UC-L, CM-38, and CH-K5, was clearly built in the fall of 1739 (Vivian 1960) with both freshly cut and stockpiled timbers (LTRR files). It contains the largest number of pre-1750 cutting dates of any site outside the Largo-Gobernador area, but other sites such as CM-18, also were contemporaneous with Dinétah area pueblitos. One site on Chacra Mesa that has not been tree-ring dated, the Doll House Site (29SJ1613), contains a ceramic assemblage typical of eighteenth-century pueblitos and was probably occupied prior to 1750 (Brugge 1986). In addition to those on Chacra and

Figure 46. Pueblito site NLC-LLC-UO-DD in Upper Oraibi Wash

Big Bead Mesas, undated pueblitos that were probably occupied at the same time as or slightly later than those in Dinétah are located near Chaco Wash (Reher 1977:54) and on Lobo Mesa (Winter et al. 1993:81).

Navajo sites in areas to the west tend to date later than 1755 (Bannister, Dean, and Gell 1966; Bannister, Hannah, and Robinson 1966), but there is still evidence that some were occupied prior to that time (Table 23). Twelve sites contain pre-1755 cutting dates, but most cannot be analyzed because of the lack of provenience and assemblage information. Gilpin (1996) discusses probable pre-1800 sites near Canyon de Chelly, on the Defiance Plateau, in the Black Creek Valley, and on Black Mesa. Although most of the pueblitos and fortified crags near Canyon de Chelly (James 1976) postdate 1755, several forked-pole hogans sites, including the Lone Tule Wash and the Sheep Dip Creek sites, contain ceramic assemblages suggesting pre-1770s occupations (Gilpin 1996:185–86). Interestingly, pueblito site NLC-LLC-UO-DD in Upper Oraibi Wash, appears very similar to boulder-top pueblitos in the Dinétah (Figure 46). Tree-ring samples from the site yielded only one near-cutting date 1790+G, and the site description is somewhat meager.

Navajo and Hopi oral traditions indicate that Hopis fleeing the destruction of Awatovi in 1700 joined an extant Navajo community in

Canyon de Chelly; oral traditions also indicate that the Navajo Taachii'nii Clan is descended from Awatovi survivors who settled at Tachii, a spring at the eastern edge of Black Mesa (Gilpin 1996:171). Navajo oral traditions also suggest a pre-1700 occupation of the Grand Canyon area (Begay and Roberts 1996). Thus, it is highly probable that at least some sites west of the Chuska Mountains were occupied prior to 1755 and the abandonment of Dinétah.

The "pulls" that attracted the Navajos to various areas outside the Largo-Gobernador area have yet to be identified. Climate, which was "improving" in the mid-1750s in the Largo-Gobernador area, was also improving in the Big Bead Mesa, Cebolleta Mesa, Canyon de Chelly, and Black Mesa areas (Dean and Robinson 1977). Thus, there was no environmental gradient (cf. Ahlstrom et al. 1995) that made any of these areas more attractive, from a climatic standpoint, than the Largo-Gobernador area.

The idea that the Navajo were seeking better pastures for their expanding herds (Haskell 1975) has not been seriously tested. Historical references (Hill 1940) indicate that a few small herds were present, but their importance in the Navajo economy is still unknown. Bailey and Bailey (1986) indicate that sheep herding did not become important until after the abandonment of Dinétah. Although Marshall (1991:249) suggests that sheep corrals and lamb pens are present near some pueblitos, surveys around other pueblitos (Dykeman and Wharton 1996; Towner and Johnson 1998) have failed to locate any features related to pastoralism. There is simply no solid evidence that sheep and the desire for good pastures were a pull factor in the abandonment. There is more evidence that sheep herding grew *after* the Navajo left Dinétah, but that cannot be cited as a pull factor in the decision to leave the area.

Another factor that has not been examined, but that deserves close scrutiny, is the possibility of overhunting. There are references throughout the historical period to the Navajo trading buckskins to the pueblos (cf. Hammond and Rey 1966). The faunal remains at pueblito sites (Brown and Brown 1995) routinely include deer bones, but the overall numbers are relatively small. There is archaeological evidence of antelope procurement in substantial quantities (Reynolds et al. 1984). It remains to be seen whether the overexploitation of deer and pronghorn contributed to the Navajo decision to emigrate, but this may be a fruitful line of investigation in the future. Historical records, including procurement documents such as supply orders or bills of sale, may be particularly important in this regard.

If Ute and Comanche raiding was the primary push factor in the emigration from Dinétah, security from such raids may have been the primary pull factor. The movement south toward Big Bead Mesa and the nearby Spanish settlements suggests that protection from raids was important. Indeed, some Navajos requested Spanish protection from Ute predation (Reeve 1959), and such requests may have been a factor in the establishment of missions at Cebolleta and Encinal in 1749 (Hackett 1937). Spanish military force may have been a factor in the decision for those Navajos who moved south, but cannot account for movements to the west. Brugge (1986:142–43) suggests that while Dinétah was occupied, the Chacra Mesa area was unattractive to raiders. One might infer that distance from the Dinétah, and the bases of raiders, was a factor in deciding where to move—provided there was an extant Navajo population willing to incorporate the migrants. Distance, however, may not have provided enough security, and other conflicts may have developed. That security was important is evidenced by the construction of several pueblitos west of the Chuskas, such as Kinnazinde, Ganado Crag, and Nazlini Pueblito in the early 1760s (Gilpin 1996). Battles with the Ute have also been documented as far west as Bidahochi in northeast Arizona (K. Kelley, personal communication). Thus, whereas security from Ute raiding seems adequate as a pull factor into the Big Bead Mesa and Cebolleta Mesa areas, it remains suspect for areas west of the mountains.

SUMMARY

Climate has certainly had an impact on the Navajo entry into, occupation, and abandonment of the Dinétah. The relationships between climate and these processes, however, are sometimes not what could be expected. The archaeological data are still too sparse to unequivocally determine when the first Navajos moved into the Dinétah; we do know that when some Navajos were living in forked-pole hogans near the San Juan River, climate was relatively favorable for agriculture (Hancock 1997; Sesler and Hovezak 1996). We also know that climate, or at least precipitation, fluctuated during the occupation of the pueblito sites. The positive correlation between tree-cutting, pueblito construction, and above average precipitation seems to be more related to Ute raiding than to environmental stress on the Navajos. I infer that it was Ute security in above-average years that prompted them to attempt slave raids against the Navajo, who responded by building boulder-top pueblitos as a defensive measure.

Climate appears to have played a minor role, if any, in the depopulation of the Dinétah. The factors driving the Navajo emigration, therefore, are unlike those postulated for the migrations of earlier occupants of the Colorado Plateau (Ahlstrom et al. 1995). The Navajo migration, however, was similar in that it was a long, complex process, not a sudden catastrophe precipitated by an environmental crisis.

CHAPTER 6

A New Model of Navajo Ethnogenesis

> The significant question, in my mind, is not whether there has been Puebloan introgression, but when it took place, the rate of incorporation, and the conditions under which it took place (Brugge 1996:256).

The dendrochronology of the pueblitos and earlier forked-pole hogans in the Largo-Gobernador area has profound implications for reconstructing early Navajo history and development. Previous models of when the Athapaskans/Navajos entered the area, when, how, why and by whom the pueblitos were built, the nature of the relationships between masonry structures and forked-pole hogans on pueblito sites, the degree of mobility of the pueblito site occupants, and the rate of abandonment of the area all must be reevaluated in light of these data. The reevaluation presented here proposes important changes in the way we interpret Navajo culture history and development, as well as shows the relevance of these data for broader issues of culture processes that may not be as obvious in fully prehistoric contexts.

Two generalized models of the Athapaskan entry into the Southwest and subsequent development of Navajo culture have provided much of the theoretical basis for interpreting Navajo archaeological materials (Towner 1997). These models of Navajo origins and ethnogenesis, which I term the "Late Entry, High Plains–Refugee Hypothesis Model" and the "Early Entry, Mountain Route–Refugee Hypothesis Model" are summarized below and shown in Figures 47 and 48. I should emphasize that no single individual has espoused all the positions indicated. Indeed, there are proponents and opponents of each position listed, and even antagonists on one issue may agree with one another on a different issue. Thus, various combinations of these models (and others) can be constructed. Nevertheless, these models, although they differ in the development of Navajo culture prior to the Pueblo Revolt, both posit significant culture change as a result of immigration by a large number of Puebloans during the pueblito site occupations. By demonstrating that the supposed influx of Puebloan refugees did not occur, at least in the ways that scholars have generally assumed, the data in this study provide the opportunity to reinterpret Navajo cultural development in ways more consistent with the dendrochronological, archaeological, written historical, and oral historical data.

The Late Entry, High Plains Refugee Hypothesis Model

This is essentially a cladistic model (Moore 1994) of Athapaskans entering the Southwest as a single entity and differentiating into Navajo and Apache groups after their arrival in their known historic ranges. It posits that the major differences between Navajos and other Athapaskans are the result of long-term interaction with and incorporation of significant numbers of Puebloans into Navajo society. The sequence of events relevant to this development are

1. Athapaskan speakers residing in the Black Hills expanded southward as pedestrian bison hunters ca. A.D. 1450, entered the Southwest via the High Plains ca. A.D. 1500 and resided on the plains of eastern New Mexico where they were known as "Querechos" (Keur 1941; Wilcox 1981).
2. Querecho/Athapaskans filtered into areas surrounding the existing pueblos between A.D. 1500–1650 and provided wild resources to the Puebloans in a mutualistic trade relationship. At the same time, the Athapaskans began to differentiate into Navajo and Apache groups (Wilcox 1981).
3. From A.D. 1640–1710, Navajo groups occupied the Piedra Lumbre area near Abiquiu, New Mexico, and began herding sheep acquired from the Pueblos; a few Navajos moved west of the continental divide into the upper San Juan River basin (Schaafsma 1975, 1976, 1978, 1996, 2002).

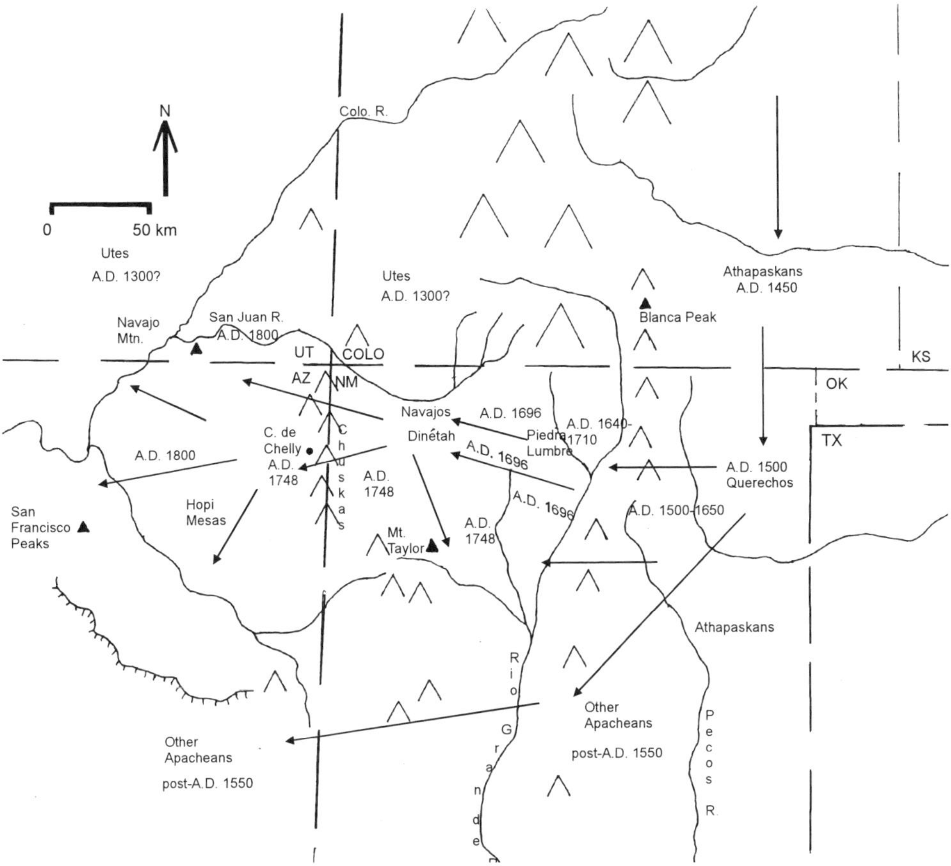

Figure 47. The "Late Entry, High Plains Refugee Hypothesis" Model

4. After the Spanish Reconquest of New Mexico (ca. A.D. 1692–1696), massive numbers of Puebloans, particularly from Jemez and Santa Clara, emigrated from the Rio Grande and took refuge with the Navajo in the Dinétah area (The Refugee Hypothesis) (Bailey and Bailey 1986; Forbes 1960; Keur 1944; Kidder 1920; McNitt 1972).

5. During the Gobernador Phase, ca. 1700–1775, Puebloan refugees, who may have constituted as much as 25 percent of the "Navajo" population (Bailey 1964), introduced corn agriculture, painted ceramics, matrilineal clans, weaving, etc., to Athapaskans (Hester 1962a), and the larger group developed into the Navajo (Bailey and Bailey 1986).

6. In 1748, drought and Ute raiding forced the abandonment of Dinétah and intensification of the pastoral subsistence regime. The Navajo moved to the south (Big Bead Mesa and Chacra Mesa) and west (Chuska Mountains and Chinle Valley) (Bailey and Bailey 1986; Hester 1962a; Marshall 1995; Reeve 1958).

The Early Entry, Mountain Route Refugee Hypothesis Model

This is also a cladistic model of Athapaskans entering the Southwest as a single entity and differentiating into Navajo and Apache groups after their arrival in their known historic ranges. It differs from the Late Entry model basically in terms of the timing and route of the original entry of Athapaskans and the degree of pre-Revolt Navajo-Pueblo interaction. The sequence of events relevant to this model of development are

1. Athapaskan speakers entered the Southwest via the mountains and valleys of Colorado (Amsden 1932) prior to A.D. 1500, and possibly much earlier (Hall 1944; Reed and Horn 1990). Indeed, some scholars suggested that Athapaskans were at least partially responsible for the Anasazi abandonment of much of the Colorado Plateau (Hall 1944; Reed 1941).

2. Some aspects of Navajo culture, such as plain ware ceramics and agriculture, developed in the upper San Juan area, including the Largo-Gobernador, Navajo Reservoir, and La Plata Mountain areas, during the fifteenth through seventeenth centuries (Hogan 1989; Reed and Horn 1990).

3. During the Early Gobernador Phase, ca. 1630–1720, Navajos acquired painted ceramics and the knowledge to produce them, as well as some forms of iconography and other items of material culture, through peaceful interaction with Puebloans (Brown 1996; Copeland and Rogers 1996; Reed and Reed 1996).

4. After the Spanish Reconquest (ca. A.D. 1692–1696) of New Mexico, massive numbers of Puebloans, particularly from Jemez and Santa Clara, emigrated from the Rio Grande and took refuge with the Navajo in the Dinétah area (The Refugee Hypothesis) (Bailey and Bailey 1986; Forbes 1960; Keur 1944; Kidder 1920; McNitt 1972). These refugees constituted as much as 25 percent of the "Navajo" population (Bailey 1964), and were primarily responsible for the introduction of matrilineal clans, weaving, and masonry

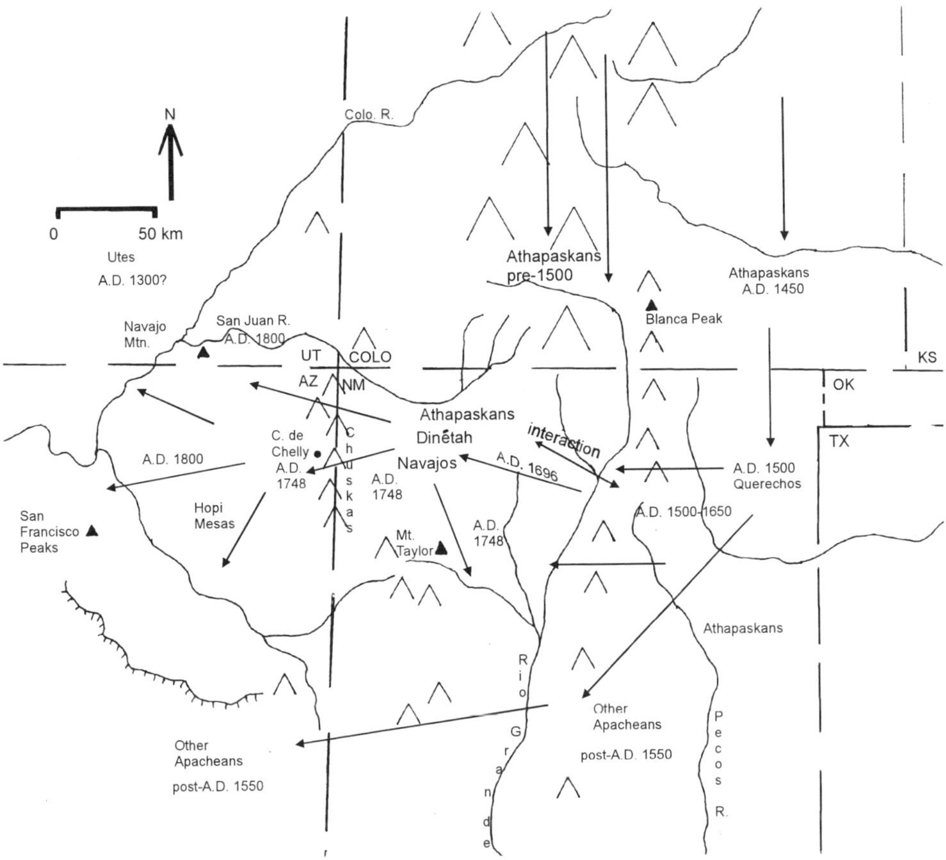

Figure 48. The "Early Entry, Mountain Route Refugee Hypothesis" Model

architecture (pueblitos) to Athapaskans (Brugge 1996). During the later part of the Gobernador Phase (1720–1775) the combined group developed into the Navajo (Bailey and Bailey 1986).

5. In 1748, drought and Ute raiding forced the abandonment of Dinétah (Schroeder 1965) and intensification of the pastoral subsistence regime. The Navajo moved to the south (Big Bead Mesa and Chacra Mesa) and west (Chuska Mountains and Chinle Valley) (Bailey and Bailey 1986; Hester 1962a; Reeve 1958; Marshall 1995). A cultural "revitalization movement" that emphasized "Athapaskan" values and the rejection of painted pottery and pueblito architecture began during or shortly after the abandonment of Dinétah (Brugge 1963).

Both models are relatively silent on the issues surrounding other Athapaskan-speakers in the Southwest. Although some efforts have been made to trace the development of the Western Apache, the models have generally relied on historical (Forbes 1960; Schroeder 1963, 1965, 1974), linguistic (Hoijer 1971; Opler 1983) and ethnographic data (Basso 1971; Goodwin 1942). This reliance is due primarily to the lack of comparable archaeological remains for the Apache (Gregory 1981), but also reflects the relatively spectacular nature of the Navajo sites in the Largo-Gobernador area. Unfortunately, I am forced to follow this trend and discuss only the Navajo.

Migration into the Dinétah

After decades of relative inattention, migration has again become a topic of interest in archaeology (Anthony 1990; Cameron 1995; Herr and Clark 1996). Thought of as a "nonexplanation" and "culture-history" by the New Archaeology of the 1970s (Cameron 1995), migration has recently been revived as a structured human behavior amenable to study by archaeologists (Anthony 1990). Two aspects of the Navajo occupation of the Dinétah are relevant to discussions of migration: immigration into the area by the initial Navajo/Athapaskan population, and the proposed immigration of the Puebloan "refugees" fleeing the Reconquest.

The immigration of the initial Athapaskan/Navajo people into the Dinétah has been characterized as (a) a slow filtering of Athapaskan-speaking "proto-Navajo" into the area in the late prehistoric and protohistoric periods (Schaafsma 1996; Wilcox 1981) and later, a massive and sudden immigration of Puebloans into the area following the Spanish Reconquest or (b) a sudden and massive immigration of Athapaskans and Puebloan refugees from the east into the area following the Reconquest (Schaafsma 1996). The differences depend on whether or not a particular researcher believes that a small population of Athapaskan/Navajo lived in the area prior to the Pueblo Revolt and was joined by a substantial number of Puebloans (Bailey 1964; Bailey and Bailey 1986), or that the Navajo were living elsewhere and moved into the area with the Puebloans (Schaafsma 1981, 1996). Support for the first immigration of a small preexisting Navajo population generally has been based on negative evidence—that is, the paucity of well-dated prehistoric Navajo sites in the area and historical references to Querechos on the Plains. The second episode of immigration, (The Refugee Hypothesis) has been viewed as a "site-unit intrusion" (Hester 1962a) that fundamentally altered Athapaskan culture to create a unique Navajo culture. This hypothesis has been based primarily on historical documents, although the architecture and ceramic assemblages at pueblito sites have been used to support the historical accounts (Hogan 1991, Towner 1996).

The Immigration into the Dinétah and Development of Pre-Revolt Navajo Culture

Schaafsma (2002) has recently reiterated his support for the Late Entry, High Plains Model and its relationship to the Piedra Lumbre phase sites. Six forked-pole sites in the Dinétah area, however, clearly demonstrate a Navajo occupation of the area prior to the Pueblo Revolt (Table 24). The sixteenth-century tree-ring dates from LA 55979, LA 74723, LA 74243 and LA 72746 indicate Navajo presence near the San Juan River in the mid- to late 1500s (Hancock 1997; Sesler and Hovezak 1996; Towner 1997). LA 55979 was built in the year of Coronado's Entrada (Hancock 1997), and LA 72746 appears to have been occupied at about the same time as Oñate began colonization of the Rio Grande (Hovezak and Sesler 1995). The other sites, LA 72767 and LA 72747 were occupied in the seventeenth century. All of these sites conform to Brown's (1996) criteria for distinguishing Navajo from Ute sites and structures; all contain Dinétah Gray ceramics, forked-pole hogans with excavated and prepared floors, interior hearths, and formal site layouts. In addition, LA 55979, the earliest dated hogan, is located south of the San Juan River and by Schaafsma's (1996) own definition cannot be a Ute site.

Although it is possible that these sites represent "some wandering Navajos . . . beginning to explore and possibly live in the San Juan drainage" (Schaafsma 1996:43), I find it unlikely that archaeological research would have located the only such sites. Furthermore, potentially earlier Navajo sites, such as Grassy Canyon (Brown et al. 1992) are located farther north, near the La Plata Mountains. Although not tree-ring dated, the other chronometric data for sites such as Grassy Canyon suggest a substantial sixteenth- or seventeenth-century Navajo occupation near the La Platas (Brown 1996).

The seventeenth-century tree-ring dates, particularly cutting dates, from hogan sites LA 105483, LA 16209, LA 72747, and LA 72767 demonstrate a Navajo occupation near the confluence of the San Juan and

Table 24. Pre-Revolt Tree-ring-dated Forked-pole Hogan Sites in the Dinétah

Site	Feature Type	Sample Number	Species	Inside Date	Outside Date	Outside Symbol	Comment
LA 55979	FE 136 Hogan	NMB-382	JUN	1453+/-p	1540r	comp	
		NMB-364	JUN	1474+/-p	1540B	comp	
		NMB-363	JUN	1464p	1541r	inc	
	Surface	NMB-374	JUN	1270+/-	1394vv		
		NMB-367	JUN	1282+/-	1522+vv		
		NMB-366	JUN	1443	1541r	inc	
	FE 124 Cist	NMB-368	PNN	1465p	1541rB	inc	
		NMB-379	PNN	1381p	1496++vv		
LA 12609	FE 1 Hogan	LPM-3	JUN	996	1303++vv		dead wood (?)
		LPM-12	JUN	1458	1559vv	inc	
		LPM-5	JUN	1383+/-p	1582+vv	inc	
		LPM-20	JUN	1423+/-p	1595+vv	inc	
		LPM-14	JUN	1390+/-	1597vv	inc	no sapwood
		LPM-7	JUN	1442+/-	1597+vv	inc	
		LPM-9	JUN	1545+/-	1600+vv	inc	
		LPM-19	JUN	1417+/-	1602vv	inc	
		LPM-2	JUN	1436+/-p	1609+vv	inc	
		LPM-18	JUN	1406+/-p	1612vv	inc	
		LPM-11	JUN	1403+/-p	1615+vv	inc	no sapwood
		LPM-8	JUN	1452+/-	1620vv	inc	ca. 30 sapwood rings
		LPM-1	JUN	1421+/-	1621+vv	inc	ca. 25 sapwood rings
	FE 2 Hogan	LPM-31	JUN	692	1111vv	inc	
		LPM-38	JUN	721+/-p	1126+vv	inc	no sapwood
		LPM-37	JUN	936	1149vv	inc	
		LPM-35	JUN	960	1211+vv	inc	
		LPM-40	JUN	986	1240+vv	inc	
		LPM-28	JUN	982	1326++vv	inc	dead wood (?)
		LPM-29	JUN	1435+/-p	1544vv	inc	ca. 47 sapwood rings
		LPM-27	JUN	1328+/-	1583++vv	inc	dead wood (?)
		LPM-36	JUN	1320+/-	1609vv	inc	
		LPM-22	JUN	1321+/-p	1611vv	inc	
LA 74243	FE 1 Hogan	NMB-112	PNN	1505	1580+vv	inc	
		NMB-119	PNN	1380p	1569v	inc	
LA 72746	FE 1 Hogan	LPM-328	PNN	1453	1546+vv	inc	dead wood (?)
		LPM-329	JUN	NO DATE			
	FE 2 Hogan	LPM-333	JUN	1426p	1593++vv	inc	dead wood (?)
		LPM-334	JUN	1406p	1596+v	inc	
	FE 4 Hogan	LPM-335	PNN	1380+/-p	1518++vv	inc	dead wood (?)
		LPM-319	JUN	1306p	1543++vv	inc	dead wood (?)
		LPM-325	JUN	1273+/-p	1547++vv	inc	dead wood (?)
		LPM-337	JUN	1479	1571++vv	inc	dead wood (?)
		LPM-324	PNN	1356p	1582++vv	inc	dead wood (?)
		LPM-323	JUN	1376p	1599+rB	comp	
	FE 5 Hogan	LPM-337	JUN	1479	1571++vv		dead wood (?)
LA 72747	FE 2 Hogan	LPM-66	JUN	1253+	1550++vv		dead wood (?)
		LPM-64	JUN	1441p	1621rB	comp	same as LPM-65
		LPM-65	JUN	—	—		same as LPM-64

Table 24 (Cont). Pre-Revolt Tree-ring-dated Forked-pole Hogan Sites in the Dinétah

Site	Feature Type	Sample Number	Species	Inside Date	Outside Date	Outside Symbol	Comment
LA 72747 (Cont.)	FE 3 Hogan	LPM-76	PNN	1384p	1495+vv		
		LPM-69	PNN	1403	1546++vv		dead wood (?)
		LPM-71	PNN	1430p	1576+v	inc	
		LPM-75	JUN	1358+/-	1586vv	inc	
		LPM-67	JUN	1313+/-	1591vv	inc	
		LPM-80	JUN	1487+/-	1610vv	inc	
		LPM-73	JUN	1447p	1614+r	comp	
		LPM-70	JUN	1483p	1615r	comp	
		LPM-74	JUN	1510p	1615r	comp	
		LPM-77	JUN	1511	1618vv	inc	
		LPM-78	JUN	1502+/-	1619v	comp	
		LPM-72	PNN	—	—		same as LPM-76
	FE 4 Hogan	LPM-91	JUN	1403	1482++vv	inc	dead wood (?)
		LPM-107	JUN	1329+/-	1484vv	inc	
		LPM-92	JUN	1384	1489vv	inc	
		LPM-100	JUN	1413	1527vv	inc	
		LPM-105	JUN	1424+/-	1540vv	inc	
		LPM-103	JUN	1394+/-	1560vv	inc	
		LPM-94	JUN	1430	1562vv	inc	
		LPM-106	JUN	1331+/-	1574vv	inc	
		LPM-82	JUN	1485	1591vv	inc	
		LPM-98	JUN	1419+/-	1610vv	inc	
		LPM-83	JUN	1509	1611vv	inc	
		LPM-101	JUN	1425+/-	1616r	inc	
		LPM-97	JUN	1550+/-	1620vv	inc	
		LPM-85	JUN	1573+/-	1621r	comp	
		LPM-89	JUN	1450+/-	1622+v	comp	
		LPM-90	JUN	1466	1622r	inc	
		LPM-81	JUN	1411+/-	1622+rB	comp	
		LPM-104	JUN	1482+/-	1622vv	inc	
		LPM-102	JUN	1438+/-	1622r	comp	
LA 72767	FE 1 Hogan	LPM-137	JUN	1441P	1592++vv		dead wood (?)
		LPM-138	JUN	NO DATE			
		LPM-139	JUN	NO DATE			
	FE 2 Hogan	LPM-117	JUN	748	984++vv		dead wood (?)
		LPM-141	JUN	1539	1641vv	inc	
		LPM-132	JUN	1428+/-P	1655+vv		
		LPM-142	JUN	1420+/-P	1676+r	comp	
		LPM-121	JUN	1524P	1676+rG	inc	
		LPM-122	JUN	1580+/-P	1677rG	inc	
		LPM-128	PNN	1618P	1677B	inc	
		LPM-135	PNN	1595P	1677rB	comp	
		LPM-136	PNN	1597P	1677rG	inc	
		LPM-133	PNN	1578+/-P	1679G	comp	

Key: NMB= Tree-ring Lab Collection; LPM= La Plata Mine Collection; JUN= Juniper; PNN= Pinyon; inc= incomplete; comp= complete

Los Pinos Rivers during the 1600s (Hovezak and Sesler 1995; Sesler and Hovezak 1996). LA 72747 in particular indicates that much of Navajo culture had developed by the early 1600s. The forked-pole hogans are typical of later sites as well, and suggest that this structure type may have a long history (Hovezak and Sesler 1995). Dinétah Gray ceramics were developed either before the Navajo moved into the area or during the initial years of the occupation. The structure of these pre-1680 hogan sites suggests that nuclear family or small extended family groups were the first to settle near the San Juan (Sesler and Hovezak 1996). Subsistence data, although limited, indicate that the Navajo included hunting, gathering, and agriculture in their subsistence base; maize, beans, and squash have all been identified at these sites, as well as at some of the La Plata area sites (Farrell 1995; Hancock 1997; Sesler and Hovezak 1996).

These sixteenth- and seventeenth-century sites clearly establish a Navajo presence in the area long before the Pueblo Revolt and suggest a northern intermountain, not Plains, route for the "Athapaskan" entry into the Southwest. They also support some Navajo oral traditions that suggest a Rocky Mountain origin (Zolbrod 1984). More importantly, perhaps, they indicate that, with the exception of painted pottery and masonry architecture, much of Navajo culture had developed prior to *extensive* contact with either Puebloans or the Spaniards. That a Navajo culture distinct from that of other Athapaskans existed in the early seventeenth century is known from Father Salmerón's use of the term "Apaches del Nabaxu" (Lummis 1900). If Salmerón was indeed referring to groups living in the Chama drainage (Schaafsma 1996, 2002), the extent of the Navajo occupation is much broader than previously thought and supports some references that the Navajo were a relatively large group who occupied much of northern New Mexico. Espejo's contact with Querechos near Mt. Taylor in 1583 may also be relevant. In both March and June of that year, Espejo encountered Querechos, first as peaceful traders and later as enemies. If these Querechos were Navajos (Hammond and Rey 1966; Reeve 1956; but see Schroeder 1963), they were living in the mountains near Acoma and probably growing corn. Likewise, it is possible that the "Cocoyes" living in the mountains near Jemez in 1598 were Navajo (Schroeder 1963).

Thus, the archaeological and historical data indicate that the Navajo occupied a relatively large area in the late sixteenth and early seventeenth centuries. At a minimum, they were living near the confluence of the San Juan and Los Pinos Rivers, near Jemez, possibly in the Big Bead Mesa area, and near Mt. Taylor, probably around Cebolleta. A small community may have been present in Canyon de Chelly as well (Fall et al. 1981; Gilpin 1996). These data indicate that early Navajo culture, including plain ware ceramics, maize agriculture, hunting, gathering, and trade with the Pueblos, was relatively well developed and distinct from other Apachean groups. I believe that the Querechos seen by Coronado on the Plains were not ancestral to the Navajo, but more likely developed into Plains Apache groups who continued to hunt bison and trade with the Eastern Pueblos. Their connection with other Querechos may have been nothing more than linguistic affinity or cultural similarities in the eyes and words of the Spanish recorders.

Only additional research in those areas will be able to substantiate or refute an early Navajo presence distinct from that of other groups. The tremendous expansion of contract-related archaeological research in the Dinétah has been responsible for many of the changes in Navajo archaeology in the past decade or two. A similar effort is needed in other areas, particularly around Mt. Taylor and near Cebolleta (Carroll 1979). The chronometric underpinnings of such research, however, must include the wood-use model developed in the Dinétah to properly interpret any dates derived from Navajo structures (Towner, 1997).

Puebloan Immigration and the Pueblito Occupations

It is clear from the tree-ring and other data that the majority of pueblitos were not built by Puebloan refugees fleeing the Reconquest. Although some of the early pueblito sites in the north, including the Cabresto Mesa Tower Complex, Los Peñoles, parts of Three Corn Ruin, and storage rooms at Frances Canyon Ruin, may have been built in response to Roque de Madrid's campaigns against the Navajo, they were still built by Navajos. The complete lack of pueblito construction and paucity of evidence for tree-cutting activities between the failed revolt of 1696 and 1705 contradict the hypothesis of a large influx of people into the area. Most Puebloan refugees who fled the Rio Grande undoubtedly went to other pueblos, such as Zuni and Hopi (Hogan 1991), and those who did not probably moved to large Puebloan "refugee" sites, such as Astialakwa, Kotyiti, Boletsakwa, and Rowe Mesa (Dougherty 1980; Elliot 1982; Kulishek 1996).

Only Tapacito Ruin contains the proper temporal and architectural attributes to be labeled a "refugee" site, and it was built by Puebloan refugees who joined an extant Navajo community (Towner and Dean 1992). Tapacito Ruin is unique in another way—it is the only pueblito site with a cluster of cutting dates from a hogan that predate a cluster from the masonry structure. Even so, the difference at Tapacito is only four years. Tree-ring dates, although often noncutting or nonclustered, indicate that most hogans and masonry structures at other pueblito sites, were occupied simultaneously. The problems with dates from hogan samples appear to be due to different wood selection criteria, rather than to any temporal discontinuities in the occupations. Tapacito Ruin was probably abandoned within a few years, possibly after the August 12, 1705, battle with the Spaniards at the confluence of the Largo River and Tapacito Creek (Hendricks and Wilson 1996). Thus, the pueblito site data indicate that there was not a massive migration of Puebloans into the Largo-Gobernador area following the Pueblo Revolt or Reconquest.

In the past, I have been criticized for not taking "full advantage of ethnographic and ethnohistorical knowledge" (Brugge 1996:261). In particular, Brugge (1996:262) believes evidence of a refugee influx exists in the origins of Navajo clans. Reichard (1928:16–19) lists nineteen Navajo clans deriving from pueblos or from Pueblo clans, including two of probable Pueblo origin. Brugge would add another five or six to Reichard's list, and discount one that she includes. In any case, approximately 40 percent of all Navajo clans claim Pueblo origins (Reichard 1928:22–25; Williams 1970:Table 6). Kelley (1993) suggests that as many as 12, and possibly 16, of the 26 clans living along the Transwestern Pipeline project areas may be Puebloan in origin.

Ethnohistorical data include Navajo oral traditions that describe the origin and growth of the tribe from its place of origin in the San Juan Mountains through times of wandering and, finally, settlement along the San Juan River. Few Navajo oral traditions relate directly to the pueblitos or the time period when they were occupied; oral traditions do, however, relate to the formation of the clans and tribe and provide some clues concerning the timing of various events. Some clans, by tradition and history, must have formed after European contact, for example, the *Naakaiidine'é* (Mexican Clan). Kelley (1993) postulates that these people may have been Genizaros with an Aztec tradition; a more likely source, however, is the relatively nearby Genizaro community of Abiquiu which was composed of people from Santa Clara and the Hopi villages (Cordell and Yannie 1991:106). (Genizaro was a particularly New Mexican term denoting Spanish-speaking Christian Indians who had become displaced from their pueblos of birth and were often captives who had been ransomed by Spanish settlers [Cordell and Yannie 1991:101]). Other clans, including some derived from non-Navajo groups, may have formed prior to contact and certainly prior to the Pueblo Revolt; few, however, trace their ancestry directly to refugee status as a result of the Reconquest.

One well-known tradition involves the San Rafael Canyon area, a Navajo sacred place (Kelley and Francis 1994; Van Valkenburgh 1974). The Navajo name for the canyon is *Taanez'a*, [Poles (Fence) Strung Out], and it is the place of origin of the *Ta'neeszawnii* [Poles (Fence) Strung Out People] clan (Van Valkenburgh 1974:150). Matthews (1897:142) states,

> the strangers encamped at a place called *Tha'nez'ni*, among the Scattered Hills. Soon after this a party came and joined the Navajo (who were then living on the San Juan River) making a new clan, which was called *Tha'neza'ni*. The strangers said that they were descendants of the Mirage People. The remains of their old huts are still to be seen at *Tha'neza*.

Likewise, Zolbrod (1984:299) indicates, "Those wanderers were invited to come and join the Navajo, which they gladly did. And soon there was another clan added to the tribe. It became known as *Ta'neeszah dine'é*. . . . These newcomers said that they were descendants of *Hadahoniye' dine'é* the Mirage People." Austin (1972:6) indicates these are the Tangle People.

Some clans were formed by various "strangers" who joined the Navajo freely. For example, the *Ts'ayisk'idnii* (Sagebrush Hill) Clan, was apparently derived from Ute visitors when the Navajo lived along the San Juan (Matthews 1897:144); likewise, the *Noóóda'ídine'é* (Ute) Clan joined the Navajo freely. Other clans originated by non-Navajos who were neither captives nor refugees include the *Tse'zhiin nidii'áhĭ* (Slanted Lava Spire), *Deeshchii'nii* (Red Streak), *To'baazhní'ázhi* (Two Came for Water), *Naasht'ézhí dine'é* (Zuni), *and Dilzhééhé* (Tonto) clans. Indeed, Kelley (1993:B-1) suggests that "both modern Puebloans and modern Navajos have many ancestors in common, including the prehistoric inhabitants of the Colorado Plateau."

Captives, not refugees, may have been able to exert considerable influence in some cases (Brugge 1985:137–38; 171–75; 1993b). Navajo tradition indicates that captive girls founded the *Áshiihí*, *Kĭnlĭtsi*, and *Tlizilani* clans. Brugge (1966c) traces the origins of both the *Nakaidiné* and *Toyahedliinii* clans to two Spanish girls captured *during* the Pueblo Revolt of

1680. Likewise, a Jemez girl captured *before* the Reconquest by a member of the *Tl'aschii* Clan was the founding ancestress of the *Ma'iidesgizhnii* (Coyote Pass) Clan (Brugge 1966b); *later refugees* from the destruction of Astialakwa (July 24–August 8, 1694) joined the clan (Brugge 1966a). Towner and Dean (1992) have suggested that these refugees may have begun construction of Tapacito Ruin within a few weeks of their escape from the Spaniards. Brugge (1966a, 1966b, 1966c) suggests that people from Zia and Santa Ana who were allied with the Jemez joined the Navajo in Dinétah and formed the *Tl'izí-lízhiní* and *Tl'ogí* clans. The Navajo *Táchii'nii* Clan is said to be descended from some of the survivors of the destruction of Awatovi who settled at Tachii, a spring at the eastern edge of Black Mesa (Brugge 1974:29; Gilpin 1996). Although technically "refugees," the *Táchii'nii* were not fleeing the Spaniards, but other Hopis.

The archaeological data from the pueblito sites and ethnohistorical data concerning clan origins suggest little immediate post-Revolt or post-Reconquest immigration. The ethnohistorical data indicate that only two clans, the *Tl'izí-lízhiní* and *Tl'ogí*, originated with refugees fleeing the Reconquest of 1692 or later events in the Rio Grande Valley, although the *Ma'iidesgizhnii* clan was apparently augmented by refugees from Jemez. Although two clans may have originated, and another may have been enlarged by refugees, the actual influx of people may not have been great. Indeed, Hogan (1992) suggests a few hundred individuals at most, and I have suggested even fewer (Towner 1996).

There has undoubtedly been abundant interaction between Navajos and Puebloans, and the Navajos have incorporated many aspects of Puebloan culture into their own. I agree with Brugge (1996:262) that not "all Puebloan infusion need be the result of the Reconquest." Indeed, I believe most was not. The prevailing model of post-Revolt Navajo cultural development, however, has been one of a massive influx of Puebloan refugees as a direct result of the Reconquest and their subsequent domination of Dinétah Navajo culture for several decades (Bailey and Bailey 1986; Hester 1962a). This is a far cry from the idea that "Navajo acceptance of Puebloan and Hispanic traits followed circuitous paths" (Brugge 1996:264) and that incorporation of Puebloans into Navajo society was a long-term process based on types of interaction developed over decades, if not centuries (Hogan 1991; Reed and Reed 1992). It is the timing, rate, and processes of such infusion that should be of interest to archaeologists and anthropologists. Processes such as mutualistic trade relationships

(Gregg 1991; Spielmann 1982) deserve closer scrutiny than they have received in the past.

The Pueblito Sites Occupations: Ceramic Data

In the past, the concept of a refugee influx has been supported by two other data sets: the apparently sudden and widespread appearance of Gobernador Polychrome ceramics, and the concomitant appearance and distribution of pueblitos. If there was no influx of refugees, these data must be explained in other ways.

Recent research on the development of Gobernador Polychrome indicates that it was developed by forms of Navajo-Pueblo interaction that were not a result of a massive post-Revolt refugee immigration (Reed 1995; Reed and Reed 1992). Reed and Reed (1996) suggest that it was manufactured prior to the Pueblo Revolt, and possibly as early as 1630; Marshall (1995:203) would push the starting date back to 1625. Although Reed and Reed (1996:103) suggest that five sites represent evidence of pre-1700 use or manufacture of Gobernador Polychrome, the evidence is compelling only at LA 72769 occupied in 1677, and LA 16209, occupied sometime after 1621. Nevertheless, it is clear that Gobernador Polychrome was being produced and used in the Dinétah prior to any purported influx of refugees fleeing the Reconquest. Reed and Reed (1992, 1996) also suggest that Gobernador Polychrome represents a uniquely Navajo product that combines stylistic elements of Hopi yellow ware, Rio Grande glaze wares, and Pueblo matte-paint ceramics from Zuni, Acoma, and the Rio Grande pueblos. Given the similarities between Gobernador Polychrome and Hopi yellow wares, and the presence of Rio Grande glaze ware elements as well (Reed and Reed 1992; Reed 1995), I believe it represents the incorporation of people and/or ideas from several different traditions. An intriguing possibility that deserves closer scrutiny is the possibility that Hopi Genizaros from Abiquiu had an influence on the development of Gobernador Polychrome. Although not mentioned in historical documents until 1715 (Swadesh 1966), the Spanish founders of Abiquiu were reinstated on "ancestral lands" after the Reconquest. If the families lived in the Rio Chama Valley near Abiquiu prior to the Revolt, it is possible that there may have been Hopi Genizaros there as well.

The distribution of Gobernador Polychrome also deserves mention. Despite its apparent association with refugees and "decline" in the mid- to late 1700s, it occurs in virtually identical proportions in the three areas where quantifiable data are available. In the Navajo Reservoir District (Hester and Shiner 1963), at Big Bead Mesa (Keur 1941), and at pueblito sites (Marshall 1991, 1995) the ceramic assemblages typically consist of approximately 80 percent Dinétah Gray, 12 percent Gobernador Polychrome, and 8 percent Puebloan tradewares (Towner 1994). If refugees were the makers of or the "inspiration" for Gobernador Polychrome, their influence was apparently equal in all of these areas. Past researchers, however, have suggested a refugee influx only in the Largo-Gobernador area. In addition, Gobernador Polychrome occurs in similar proportions on sites west of the Chuska Mountains as far west as Black Mesa, Arizona (Gilpin 1996). Finally, there has been some suggestion that Gobernador Polychrome represents an "elite" ceramic type. Again, where quantifiable data are available, it occurs in ceramic assemblages associated with a wide variety of site types and is not restricted to pueblitos or other large sites (Towner and Johnson 1998). Thus, if Gobernador Polychrome was associated only with refugees, they were apparently living in pueblitos, single-hogan sites, and multihogan clusters and using sweatlodges, ramadas, and other small, limited-use sites. It is much more probable that the tradewares found on pueblito sites provided models for Navajo production of Gobernador Polychrome.

The Pueblito Site Occupations: The Structures

Three implicit assumptions have guided much of the research concerning the eighteenth-century Navajo occupation of pueblito sites. First, for many years it was believed that the pueblitos housed Puebloan refugees and the associated forked-pole hogans were Athapaskan/Navajo dwellings (Kidder 1920). Second, because of the use of a "phase" system, many researchers assumed that, with the possible exception of Tapacito Ruin, most pueblito sites were simultaneously occupied. This assumption has had important implications for reconstructions of Navajo demography and culture change. Finally, a corollary assumption has been that the pueblito sites were long-term occupations inhabited for the majority of the Gobernador phase (1690–1775).

If the first assumption were true, and refugees were joining extant Navajo communities, the forked-pole hogans on the sites should consistently predate the construction of masonry structures. This is simply not the case. Although the tree-ring dates from the hogans are few and difficult to interpret, it is apparent that the two structure types are contemporaneous and, with the possible exception of Tapacito Ruin, forked-pole hogans and pueblitos were usually built at the same time. This structure contemporaneity suggests that entire integrated communities, probably nuclear or extended-family groups, built and occupied all the structures on a site. There is no evidence, except at Tapacito Ruin, of a masonry structure being constructed in the midst of an extant community.

The tree-ring data obviously disprove the assumption of long-term occupation. Even though the majority of pueblitos were built between 1710 and 1754, some sites were abandoned prior to the 1750s, and others were not built until the 1750s. Of the sites with defined construction episodes, never was there construction at more than eight sites in the same five-year period, and the average number of sites with construction during a five-year period was five. Thus, some pueblitos were built and abandoned before construction began on others. Interpueblito communities undoubtedly existed (see Chapter 4), but to assume that all pueblitos were occupied simultaneously ignores the tree-ring data and seriously distorts the nature of the eighteenth-century Navajo occupation of the Dinétah.

Despite the presence of habitation features and large middens, the tree-ring data also indicate that most pueblitos, even large structures, were used for relatively short periods—although some were built, remodeled, abandoned, and reoccupied. One of the largest pueblitos, Old Fort Ruin, was constructed, occupied, and abandoned in less than ten years. Some of the smaller pueblitos, such as Kin Wólá Jíní and Simon Canyon, show no evidence of having been occupied for even as much as five years. I believe that many of the pueblito sites, especially those close to each other, may have been used by the same group of people; whether such a group consisted of a clan, lineage, nuclear or extended family, or "outfit" cannot be determined with the current data. The concept that several pueblito sites were occupied by the same group at different times, however, is consistent with research on later Navajo settlement patterns in several areas (Gilpin 1993a, 1993b; Oswald 1993).

This pattern of intersite mobility can be seen in many other Navajo settlements, most notably those documented after the transition to a pastoral economy. Oswald (1993) and Blomberg (1983) document settlement mobility in the Black Mesa, Arizona, area. One of Oswald's (1993) informants, Joan, lists 38 "sites," including 48 different habitation structures, such as hogans or log cabins, in her traditional grazing area of approximately 20 square miles on northern Black Mesa. Joan's family "established a settlement, used it for a while, abandoned it, and built another settlement close by" (Oswald 1993:162). Another informant, Marge, lists 40 "sites" with 16 "habitation" structures; the smaller number of habitation structures, however, is a result of her family often spending the winter at "summer" sites. "When the family no longer wanted to use a site, they established a new one a short distance away. . . . Some family members would start building structures on the new settlement, while other relatives continued to use the older settlement. . . . Eventually, everyone would use the new settlement" (Oswald 1993:148).

Although much of this twentieth-century mobility was influenced by the availability of forage for herding animals, other factors were also important. Field locations often determined the location of some site types, but the availability of reliable water, an adequate supply of firewood, natural shelter from winter storms, and other factors were important as well. Usually, microenvironmental differences, not large-scale changes in environment or subsistence practices, precipitated a move. A similar system may have operated in the Largo-Gobernador area.

These data have important implications for estimating the mobility of the eighteenth-century Navajo, the size of the Dinétah Navajo population, and the abandonment of the area. They indicate that the pueblito sites were occupied for relatively brief periods of time and some were sporadically reoccupied. This interpretation is very different than the past assumptions that the sites were continuously occupied for decades and expanded through population growth and immigration. Despite the progress in assessing the pueblito site occupations, there are several aspects of them that I cannot assess with the tree-ring data. First, I cannot determine whether the sites were occupied seasonally or on a year-round basis. Although I suspect some sites were year-round residences and others were not (cf. Brugge 1986), additional data are necessary before site seasonality can be addressed. Many of the pueblitos contain fireplaces, multiple rooms, and middens that suggest permanent residence; whether such residence was year-round or not, however, is still unknown. Architectural features that might indicate use during specific seasons, such as granaries or lambing pens, are generally absent or not temporally associated with the Navajo occupations. Second, without much more detailed archaeological data, I cannot ascertain whether the pueblitos were habitation structures (either permanent or temporary) or storage structures also used as temporary refuges (cf. Jacobson et al. 1992). Again, I suspect that both functions are present in the site sample and that the same site may have served different functions over the course of its use-life. Most pueblitos do not indicate repeated repair episodes, however, such as might be evident if they were used as temporary refuges on multiple occasions—as described by Van Valkenburgh (1940). Finally, there is no way to detect whether specific family or extended groups used specific pueblito sites at different times of the year or in different years. Navajo oral tradition (Matthews 1897, 1994) suggests that the pueblitos in San Rafael Canyon were all occupied by members of a single clan, and Towner and Dean (1992) suggest a group of Puebloans joined an extant community at Tapacito Ruin. Gilpin (1996:185) indicates that a group of later pueblitos west of the Chuska Mountains may have all been under the control of a single leader, a survivor of the destruction of Awatovi. Thus, there is some evidence that specific groups used all the structures in a specific area. Additional research, including oral histories, should be conducted regarding the possible communities identified in Chapter 4.

In general, pueblito construction appears to have shifted geographically during the seventy-five years such sites were built, and construction at different times in different places was probably a response to different stimuli (Towner, Rogers, and Copeland 2001). It is important to remember that different pueblitos were built in different settings at different times for different reasons. The pre-1720 long-distance pueblito network and rockshelter sites suggest both a large-scale Dinétah network and individually defensive (family?) responses to the Spanish threat. Pueblitos built between ca. 1720 and 1745 in response to Ute raiding suggest that families and suprafamily groups (outfits?) utilized specific areas, but did not participate in a "pan-Dinétah" defensive network. Finally, the 1745–1755 pueblito sites indicate that some Dinétah Navajos aggregated into specific areas (e.g., San Rafael Canyon and Frances Mesa), but that much of the area was sporadically occupied by smaller groups. The "Navajo fort" community of the 1760s remains enigmatic.

It is clear from these data and other research (cf. Dykeman and Wharton 1996; Sesler, Hovezak, and Wilshusen 2000) that pueblitos comprise but a small and varied component of the Navajo occupation of Dinétah. Pueblito sites, therefore, may also represent but a small portion of the Navajo population. This hypothesis is consistent with Brugge's (1968a:36) inference that "in spite of the spectacular nature of cultural developments in the Dinétah during this period, it was not the center of the greatest Navajo population." I believe the importance of pueblitos in Navajo cultural development has been vastly overstated for two reasons. First, pueblitos are impressive structures, especially when compared to forked-pole hogans, and have attracted the attention of archaeologists. Other areas, such as Cebolleta, contain large numbers of forked-pole hogan sites (Carroll 1979) that are associated with Dinétah Gray and some painted Navajo ceramics, yet have received little attention from archaeologists. If the Dinétah Navajos had not constructed pueblitos, I doubt the Refugee Hypothesis would have developed and would not have become the dominant model of Navajo cultural development. The low proportion of Puebloan ceramics on these sites, similar to proportions at Big Bead Mesa and in the Navajo Reservoir District, would have been viewed as evidence of trade or occasional peaceful interaction, not as evidence of a massive immigration of people.

Second, pueblitos have been important to Navajo archaeological research because they have provided the majority of tree-ring dates from early Navajo sites, something that has also attracted the attention of archaeologists and historians. This is an archaeological phenomenon, however, not necessarily a cultural phenomenon. Simply put, the good preservation of timbers in pueblitos, combined with the selection criteria used by the builders, has ensured that the timbers in masonry structures produce many high-quality tree-ring dates. Forked-pole hogans, on the other hand, have been sampled far less frequently by archaeologists and have produced far fewer and less satisfactory dates because of the criteria used in selecting timbers and their poor preservation after site abandonment. I believe that when archaeologists start intensively sampling forked-pole hogans in other areas, using the wood-use model in Towner (1997), they will generate many pre-1750 dates in areas thought not to have been occupied until after the abandonment of Dinétah.

A NEW MODEL OF NAVAJO ETHNOGENESIS

The failure of the "Refugee Hypothesis" to explain the nature of Navajo cultural development in the late sixteenth and seventeenth centuries and the variability of pueblitos does not mean there was no introgression of Puebloan ideas or people into Navajo society. Quite the contrary, there is ample evidence that Navajos and Puebloans have interacted in a variety of ways, both as adversaries and allies in times of conflict, and as "merchants" and "customers" during episodes of peaceful exchange. It is these forms of interaction during at least the past 500 years that have influenced both cultures and that deserve additional study.

Changing our perceptions of Navajo cultural development during the

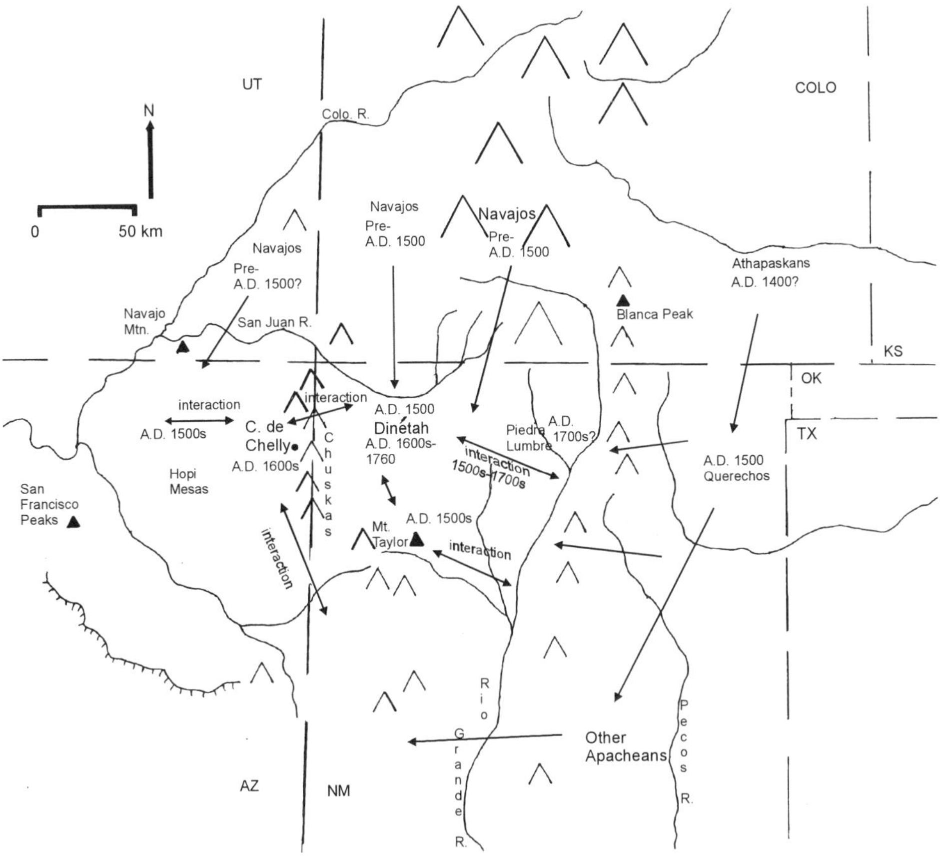

Figure 49. A new model of Navajo ethnogenesis

pueblito site occupations requires examining the data for the Navajo prior to that time as well. I believe that a well-developed Navajo culture, distinct from that of other Athapaskans, existed in the middle of the sixteenth century, and possibly earlier. These early Navajos lived across a broad area of north-central New Mexico and possibly in Arizona as well (Figure 49). The Navajos apparently entered the area from the north, probably via the mountains and valleys of Colorado, but the exact nature of their culture outside New Mexico has yet to be identified. Likewise, the timing of their entry remains open to question, although their presence as a distinct group in the mid-1500s suggests that at least a few hundred years were necessary for them to differentiate from other Athapaskans. This differentiation also suggests that the Querechos seen by Coronado in 1541 were ancestral to non-Navajo Athapaskan groups who relied on bison hunting as a mode of subsistence. If the Navajo as a distinct group were living from the La Plata Mountains and adjacent areas in the north, south to the Mt. Taylor–Cebolleta–Acoma area, and as far west as the Chuskas in the late 1500s, their population was probably larger than some estimates (ca. 7,000; Hester 1962a), but not as large as the Benavides estimate (ca. 200,000) (Brugge 1994). This model does not necessarily exclude the Piedra Lumbre remains (Schaafsma 1979, 2002) from being Athapaskan, but it precludes them from being Navajo.

Because the Navajos were relatively widespread by the late 1500s, incorporation of other people into Navajo society was both easier and more complex. Certainly there was some post-Revolt immigration in the Dinétah, most notably at Tapacito Ruin. Other Puebloan immigrants probably joined Navajo groups living nearer the Rio Grande at Cebolleta and Big Bead Mesa. Additional incorporation of Puebloans into Navajo groups occurred west of the Chuskas with the immigration of displaced Hopis. Marshall's (1979) identification of "Tecolote Red Ware" in the Cebolleta area as a product of immigrants from the Laguna-Acoma district suggests that Navajos outside the Largo-Gobernador area incorporated Puebloans into their own groups during the eighteenth century, and probably earlier as well. I believe that much of the introgression involved peaceful interaction between these Navajos and nearby Puebloans with whom they had established long-term trading relationships (Reed and Reed 1992).

Both Spanish and Puebloan captives have already been mentioned as founders of various clans. The growth of these clans depended not only on natural reproduction, but also on the acceptance of newcomers who had some relationship, however tenuous, with the established clan members. The growth of the *Ma'iidesgizhnii* clan represents this type of growth, albeit the newcomers were refugees. During the occupation of the pueblitos, there was certainly some growth of individual settlements and of the Navajo as a cultural group; however, it was not as great as has been assumed in the past.

CONCLUSIONS

The pueblitos of the Largo-Gobernador area have played a critical role in interpreting Navajo cultural development. These stone and wood structures, and their associated artifacts, guided interpretations of eighteenth-century Navajo cultural development for much of the twentieth century. The cultural elaboration seen in the Largo-Gobernador area during the

1700s has been interpreted as a demographic shift, primarily as a result of the immigration of Puebloan refugees from the Spanish Reconquest of New Mexico in 1692. Evidence for this "cultural florescence" has consisted of historical documents suggesting a refugee immigration, the appearance of Gobernador Polychrome ceramic technology, the presence of numerous masonry structures thought to represent refugee construction techniques, and, to a lesser extent, rock art styles (Schaafsma 1963). Hogan (1991) has demonstrated that the historical documents are ambiguous regarding the number of refugees who emigrated to the Navajo country. He suggests a few hundred individuals at most, and I have suggested fewer (Towner 1996). Gobernador Polychrome, once thought to represent a refugee influx, was probably produced several decades prior to the Pueblo Revolt and appears to be a mixture of ceramic styles and technologies. Its distribution in the eighteenth century, from the Largo-Gobernador area to west of the Chuska Mountains, also suggests that it was not limited to refugee production or "inspiration." Likewise rock art styles, once thought to represent Navajo acceptance of refugee iconography, are present across broad areas of northwestern New Mexico and into northeastern Arizona possibly prior to the Pueblo Revolt, and indicate a wide-ranging, yet relatively cohesive cultural group (Copeland and Rogers 1996; Gilpin 1996).

Finally, most pueblitos—another proposed hallmark of refugee influence—were not built until more than fifteen years after the supposed influx. The temporal, topographic, and architectural variability of pueblito sites indicates that they were not refugee-built structures designed for a single purpose, but were multi-faceted Navajo adaptations to different stimuli. Furthermore, the number of contemporaneous pueblitos is relatively small, and few were occupied continuously for more than a few years. Pueblitos preserve well, however, and have yielded picturesque photographs (Roessel 1983) and tree-ring dates, both valuable commodities for archaeologists. It is apparent from this study that I believe that the population living in pueblitos was relatively small and was probably not representative of Navajos living in other areas to the south and west. The Dinétah Navajos were simply responding to much more intensive and serious Ute and Comanche raiding by building stone structures and moving between different stone structures more frequently. This generally agrees with Brugge's (1968a) contention that the Largo-Gobernador area was not the center of Navajo population. Thus, the cultural elaboration seen at the pueblitos sites, and throughout the Largo-Gobernador area, should be viewed as a relatively temporary response by a small group of Navajos to threats more serious than those faced by Navajos living in other areas. Many of the cultural traits associated with this time period had already been developed, both within and outside the area, prior to the emigration of the Largo-Gobernador Navajos from the Dinétah. When they finally depopulated the Largo-Gobernador because of Ute/Comanche warfare, they were quickly assimilated into other areas with preexisting Navajo populations that had undergone their own development in the early eighteenth century. Pueblitos continued to be adaptive for a brief period, especially west of the Chuska Mountains, but were eventually abandoned. Pueblitos probably fell into disfavor because Ute raiding declined, warfare with the more heavily armed Utes and a new generation of Spaniards increased, and pueblitos were ill-suited to a pastoral settlement pattern. Forked-pole hogans, on the other hand, continued to be used precisely because they were already adapted to a relatively mobile settlement pattern and could be hidden, abandoned, or both during times of strife with large military forces.

Appendix. Pueblito Site Tree-ring Dates Used in the Analysis but Not Presented in the Text.

Site Name Number	Sample Number	Species	Provenience	Inside Date	Outside Date	Terminal Ring	Tool Marks	Comment
42 PUEBLITO	NLC-1388	PNN	Floor fill RM 2	1635+p	1728	inc	Unknown	
	NLC-1389	PNN	Floor fill RM 2	NO DATE	—		Unknown	
	NLC-1390	PNN	Floor fill RM 2	1640+p	1727	inc	Unknown	
	NLC-1391	PNN	Floor fill RM 2	1636+p	1725	inc	Unknown	
	NLC-1392	PNN	Door lintel (sealed)	1611p	1689	inc	Unknown	
	NLC-1393	PNN	Door lintel (sealed)	1627p	1692	inc	Unknown	
	NLC-1394	PNN	Roof primary RM 1	1557+p	1708	inc	Unknown	
	NLC-1395	PNN	Roof primary RM 1	1633+p	1731	inc	Unknown	
	NLC-1396	PNN	Loose log north side	1599+p	1696+		Unknown	
	NLC-1397	PNN	Loose log north side	1614p	1711	inc	Unknown	
	NLC-1398	PNN	Ladder pole north side	1494+p	1719	inc	Unknown	
	NLC-1399	PNN	Loose log SW side	1614+p	1703	inc	Unknown	
	NLC-1400	PNN	Loose log SW side	1566+p	1727	inc	Unknown	
	NLC-1401	PNN	Loose log SW side	1621p	1721	inc	Unknown	
	NLC-1402	PNN	Loose log SW side	1615+p	1711	inc	Unknown	
	NLC-1403	PNN	Loose log SW side	1604+p	1710	inc	Unknown	
	NLC-1404	PNN	Hogan leaner 2	NO DATE	—		Unknown	
	NLC-1405	PNN	Hogan leaner 2	NO DATE	—		Unknown	
	NLC-1406	PNN	Hogan leaner 2	NO DATE	—		Unknown	
	NLC-1407A	PNN	Hogan leaner 2	1608p	1699G	inc	Unknown	
	NC-4	PNN	Loose log	1609p	1741vv	inc	Unknown	
ADAMS	DNT-121	PNN	Loose log	1543	1692vv	inc	MA ends, limb	weathered
CANYON	DNT-122	JUN	Loose log	1609	1736+vv	inc	MA ends	weathered
PUEBLITO	DNT-124	JUN	Loose log	NO DATE	—		MA ends	weathered
LA 55824	DNT-125	JUN	Loose log	1528	1649vv	inc	MA ends	weathered
	DNT-123	JUN	Hogan leaner Loose	1530	1630vv	inc	MA ends	weathered
ADOLPHO	NMS-9	PNN	Hogan 1	1621p	1733vv	inc	Unknown	
CANYON	NMS-10	PNN	Hogan 1	1621p	1747+v	inc	Unknown	
RUIN	NMS-4	PNN	Hogan 2	1495p	1725++vv	inc	Unknown	
LA 5665	NMS-7	JUN	Hogan 6	1489+/-p	1749+vv	inc	Unknown	

Site Name Number	Sample Number	Species	Provenience	Inside Date	Outside Date	Terminal Ring	Tool Marks	Comment
	NMS-6	JUN	Hogan 6	1529+/-p	1751v	inc	Unknown	
	NMS-5	JUN	Hogan leaner	NO DATE	—			
	NMS-11	JUN	Sweatlodge	1451p	1684+vv	inc	Unknown	
	NMS-12	JUN	Hogan 1 leaner	NO DATE	—			
	NMS-13	JUN	Ramada	NO DATE	—			
	NMS-14	JUN	Door lintel RM 1	1576p	1747rB	inc	Unknown	
BOULDER	DNT-101	JUN	Roof primary (?) RM 3	1540	1720vv	inc	MA bark, limb, shape, end	flat
FORTRESS	DNT-102	PNN	Roof primary (?) RM 3	1585	1727vv	inc	MA limb, end	
LA 55825	DNT-103	PNN	Roof primary (?) RM 3	1601	1715vv	inc	MA limb, ends	
	DNT-104	JUN	Door lintel RM 2/3	1550+	1723G	inc	MA ends	
	DNT-105	PNN	Roof primary RM 2	1639	1726vv	inc	MA limb, ends	
	DNT-106	PNN	Roof primary RM 2	1558	1726vv	inc	MA limb, ends	
	DNT-107	JUN	Door lintel RM 1/2	1564p	1727B	comp	MA ends	
	DNT-108	PNN	Roof primary RM 1	1632	1728v	inc	MA end	
	DNT-109	JUN	Hogan leaner	NO DATE	—			
	DNT-110	PNN	Hogan leaner	1509	1709+vv	inc	Broken end	
	DNT-111	JUN	Hogan leaner (?)	NO DATE	—			
CABRESTO	RG-791-1	PNN	Unknown	1489p	1705vv		Unknown	ladder at base?
MESA	RG-791-2	PP	Unknown	1591p	1712v		Unknown	"Top of site"
TOWER	RG-791-7	PP	Unknown	1612p	1713r	inc	Unknown	"Top of site"
LA 2138	RG-791-8	PNN	Unknown	—	—		Unknown	"Top of site"
	RG-791-9	JUN	Unknown	1451+p	1713v	inc	Unknown	"Top of site"
	RG-791-4	PP	Unknown	1625p	1714r	inc	Unknown	"Top of site"
	RG-791-5	PP	Unknown	1620p	1714c	inc	Unknown	"Top of site"
	RG-791-3	PNN	Loose log east slope	1559p	1713c	inc	Unknown	"Top of site"
	NAV-10	PNN	Loose log east slope	1616p	1710vv	inc	Unknown	
	RG-791-6	PNN	Loose log east slope	1628p	1714c	inc	Unknown	"Top of site"
	NLC-847	PNN	Roof primary (RM 3?)	1615+/-p	1713c		Unknown	
	NLC-849	PNN	Roof primary RM 3	1596+/-p	1713	inc	Unknown	same as RG-791-8
	NAV-9	PP	Roof primary (RM 3?)	1630p	1713r	inc	Unknown	
	NLC-848	PNN	Roof primary	1590+/-nc	1712	inc	Unknown	

CANYON	DNT-405	PNN	Roof primary RM 2	1642+	1734vv	inc	MA bark, limb, end	
VIEW	DNT-406	PNN	Roof primary RM 2	1417	1732vv	inc	MA bark, limb, shape, end	flat; fork
LA 55827	DNT-407	PNN	Roof primary RM 2	1655	1734vv	inc	MA limb	
	DNT-408	PNN	Roof primary RM 2	1573p	1724vv	inc	MA bark, limb	same as NLC-1276
	DNT-409	PNN	Loose log RM 2	1650	1734vv	inc	None	
	NLC-1275	PNN	Roof primary (?) RM 2	1627p	1734v	inc	Unknown	
	NLC-1276	PNN	Roof primary RM 2	—	—	inc	Unknown	same as DNT-408
	NLC-1273	PNN	Roof primary RM 2	1639p	1726vv	inc	Unknown	
	DNT-410	PNN	Door lintel RM 2	1638	1727vv	inc	MA end	
	DNT-411	PNN	Door lintel RM 2	1621	1727vv	inc	None	
	DNT-412	PNN	Horizontal Intramural RM 2	1710	1734vv	inc	None	
	NLC-1274	PNN	Loose log RM 2	NO DATE	—		Unknown	
	NLC-1272	PNN	Loose log RM 2 (?)	NO DATE	—		Unknown	
	DNT-413	PP	Loose log Hogan area (?)	1685	1725vv	inc	None	
	DNT-414	PNN	Loose log Hogan area (?)	NO DATE	—		None	
	DNT-415	PNN	Loose log Hogan 3 fork (?)	1611p	1721vv	inc	None	same as NLC-1264
	DNT-416	PNN	Loose log Hogan 3 leaner (?)	1366p	1601+vv	inc	None	same as NLC-1263
	DNT-417	PNN	Loose log Hogan 3 leaner (?)	—	—	inc	None	same as NLC-1263
	NLC-1263	PNN	Hogan leaner 3	—	—	inc	Unknown	same as DNT-416,417
	NLC-1264	PNN	Hogan fork 3	—	—	inc	Unknown	same as DNT-415
	NLC-1262	PNN	Roof primary (?) Loose	1652p	1727v	inc	Unknown	
	NLC-1266	PNN	Loose log NE area	1613p	1701vv	inc	Unknown	
	NLC-1267	PNN	Loose log NE area	1475p	1728vv	inc	Unknown	
	NLC-1268	PNN	Loose log NE area	1693p	1732vv	inc	Unknown	
	NLC-1270	PNN	Loose log NE area	1709p	1733vv	inc	Unknown	
	NLC-1271	PNN	Loose log SW area	1710p	1733v	inc	Unknown	
	NLC-1265	PNN	Loose log SW area	NO DATE	—		Unknown	
	NLC-1269	PNN	Loose log SW area	1637p	1728vv	inc	Unknown	
THE CITADEL	DNT-147	JUN	Door lintel entry	1454p	1573++vv	inc	MA end, limb, bark	
LA 55828	DNT-148	JUN	Door lintel entry	1613	1687v	inc		
	DNT-149	JUN	Door lintel entry	NO DATE	—		MA limb	
	DNT-150	JUN	Door lintel entry	NO DATE	—		MA limb	
	DNT-151	JUN	Door lintel entry	1543+/-	1688++B			
	DNT-152	JUN	Door lintel entry	1499+/-	1683++B	inc	MA shape	flat

Site Name Number	Sample Number	Species	Provenience	Inside Date	Outside Date	Terminal Ring	Tool Marks	Comment
	DNT-153	JUN	Other log entry	NO DATE	—			
	DNT-154	JUN	Door lintel RM 1	1569	1642v	inc		
	DNT-155	JUN	Door lintel RM 1	NO DATE	—			split
	DNT-156	JUN	Alcove roof RM 1	NO DATE	—		broken limb	
	DNT-157	JUN	Alcove roof RM 1	NO DATE	—		MA limb	
	DNT-158	JUN	Alcove roof RM 1	NO DATE	—			
	DNT-159	JUN	Roof primary RM 1	NO DATE	—		poss Stone Ax end	
	DNT-160	JUN	NW shelf pole RM 2 NW	NO DATE	—		MA limb	
	DNT-161	JUN	Door lintel RM 2	NO DATE	—		MA shape	split
	DNT-162	JUN	SW shelf pole RM 2 SW	NO DATE	—			split
CROW	DNT-112	PNN	Shelf pole RM 1	1637	1723vv	inc	MA limb, end	
CANYON	DNT-113	PNN	Ladder area (Loose)	1436	1719+G	inc	MA limb, bark, end	flat
LA 20219	DNT-114	JUN	Ladder area (Loose)	1498	1703++vv	inc	MA limb, end; broken end	
	DNT-115	JUN	Loose log Area 2	1617	1716+vv	inc	MA end, broken end	
	DNT-117	PNN	Loose log Area 2	1622p	1715+G	inc	Broken ends	
	DNT-118	PNN	Loose log Area 2	1566	1706+vv	inc	Broken ends	
	DNT-119	JUN	Upright post Area 1	NO DATE	—		MA end	
	DNT-116	JUN	Burned stump	NO DATE	—		MA cuts	
	DNT-120	PNN	Loose log	1727	1705+vv	inc	MA limb, end, broken end	
DEER	DNT-554	PNN	Room 5/6 Wall Support	NO DATE	—		None	
HOUSE	DNT-558	PNN	Roof primary? RM5	1681	1723vv	inc	MA bark, limb, ends	
LA 106574	DNT-555	PNN	Roof primary RM 3	1626	1726vv	inc	MA bark, limb	
	DNT-556	PNN	Roof primary RM 3	1651	1726v	inc	MA bark, limb	
	DNT-557	PNN	Roof primary RM 3	1638	1720vv	inc	MA limb	
	DNT-559	PNN	Door lintel RM 2/3	NO DATE			MA end	
	DNT-560	PNN	Roof secondary RM 1	1551	1682vv	inc	MA limb, ends	
	DNT-561	PNN	Roof secondary RM 1	NO DATE	—		MA shape	
	DNT-562	PNN	Roof secondary RM 1	NO DATE	—		MA shape, broken end	SB
	DNT-563	JUN	Roof secondary RM 1	NO DATE	—		Burned end	SB
	DNT-564	PNN	Roof primary RM 1	1611	1739+vv	inc	MA bark, limb, ends	
	DNT-565	PNN	Roof primary RM 1	1615	1738+vv	inc	MA bark	charred

	DNT-566	JUN	Hogan leaner	NO DATE	—		None	
	DNT-567	JUN	Hogan leaner	NO DATE	—		None	
DELGADITO	DNT-487	JUN	Roof primary RM 3	1594	1681vv	inc	MA limb	
RUIN	DNT-488	PNN	Roof primary RM 3	1565	1717+vv	inc	MA limb, shape	flat
LA 5649	DNT-489	JUN	Roof primary RM 3	NO DATE	—		MA limb	
	DNT-490	PNN	Shelf pole RM 3	1609	1714G	inc	MA limb	
	DNT-491	PNN	Loose log	1630	1710+vv	inc	MA limb, shape	flat; fork
	DNT-492	JUN	Loose log	1553+	1713G	inc	MA limb, ends	
DOOYIIT'ÍÍNÍ	NMB-244	JUN	Ax-cut stump	NO DATE	—		MA cut	
PUEBLITO	NMB-245	JUN	Ax-cut stump	NO DATE	—		MA cut	same as NMB-246
	NMB-246	JUN	Ax-cut stump	—	—		MA cut	same as NMB-245
	NMB-247	JUN	Brush structure	NO DATE	—		MA cut	
	NMB-248	JUN	Brush structure	1636	1712vv	inc	MA cut	same as NMB-249
	NMB-249	JUN	Brush structure	—	—			same as NMB-248
	NMB-250	JUN	Loose log North side	1520+/-p	1713vv	inc	MA limb	
	NMB-251	JUN	Loose log North side	1606	1719v	inc	MA limb	
DOS	DNT-589	JUN	Loose log	1603	1710+vv	inc	MA limb, end	
CERRITOS	DNT-590	JUN	Loose log	1604	1732G	inc	MA end	
LA 2136	DNT-591	JUN	Ladder pole	1633	1729+vv	inc	MA limb, end	notched
	DNT-592	PNN	Loose log	1613	1716vv	inc	MA bark, end	
	DNT-593	PNN	Loose log	1686	1732vv	inc	MA limb, broken end	
	DNT-594	PNN	Dead tree (corral?)	NO DATE	—			
	RG-792-1	PNN	Unknown (Loose)	1595p	1734r	inc		
	RG-792-2	PNN	Unknown (Loose)	1476p	1734r	inc		
	RG-792-3	PNN	Unknown (Loose)	1647p	1728v	inc		
	RG-792-4	PNN	Unknown (Loose)	1356p	1727v	inc		
	RG-792-5	PNN	Unknown (Loose)	1612p	1733r	inc		
	RG-792-6	PNN	Unknown (Loose)	1390+p	1725v	inc		
FOOTHOLD	DNT-535	PNN	Roof primary RM 1	1652	1731+vv	inc		
PUEBLITO	DNT-536	PNN	Roof primary RM 1	1635	1700vv	inc	MA end, burned end	
LA 9073	DNT-537	PNN	Roof secondary RM 1	1644p	1732+vv	inc	MA limb	
	DNT-538	PNN	Roof primary RM 1	1623	1732+vv	inc	MA limb, end	

Site Name Number	Sample Number	Species	Provenience	Inside Date	Outside Date	Terminal Ring	Tool Marks	Comment
	DNT-539	PNN	Roof primary RM 1	1620	1735+vv	inc	MA shape; root flare	
	DNT-540	PNN	Roof primary RM 1	1648	1740vv	inc	MA end	
	NLC-1346	PNN	Roof secondary (?) RM 1	1669p	1739+	inc	Unknown	
	NLC-1347	PNN	Roof secondary (?) RM 1	1605p	1737+	inc	Unknown	
	DNT-544	PNN	Roof primary RM 2	1635	1683vv	inc	MA limb	
	DNT-545	PNN	Roof primary RM 2	1649	1713vv	inc	MA limb	
	DNT-546	PNN	Roof primary RM 2	1635	1734+G	inc	MA limb	
	DNT-547	PNN	Roof primary RM 2	NO DATE	—		MA limb	
	DNT-548	PNN	Roof primary RM 2	1656	1738+vv	inc	MA limb	fork
	DNT-549	PNN	Roof primary RM 2	1643	1727vv	inc	MA limb	
	DNT-550	PNN	Roof primary RM 2	1618	1727vv	inc	MA limb	
	DNT-553	PNN	Roof primary RM 2	1651	1739vv	inc	MA limb	
	DNT-551	JUN	Door lintel RM 2	NO DATE	—		None	
	DNT-552	JUN	Door lintel RM 2	NO DATE	—		None	
	DNT-541	PNN	Loose log (RM 3?)	1630	1735+vv	inc	MA end	
	DNT-542	PNN	Roof primary RM 3(?)	NO DATE	—		MA bark, end	charred
	DNT-543	PNN	Loose log (RM 3?)	1620	1732+vv	inc	MA limb	
	NLC-1348	PNN	Hogan 3	NO DATE	—		Unknown	
	NLC-1349	PNN	Hogan 3	1620+p	1667	inc	Unknown	
	NLC-1350	PNN	Hogan 3	1621p	1710+		Unknown	
	NLC-1351	PNN	Hogan 3	1613+p	1691	inc	Unknown	
	NLC-1352	PNN	Hogan 3	1628p	1702	inc	Unknown	
	NLC-1353	POP	Hogan 3	NO DATE	—		Unknown	
	NLC-1354	PNN	Loose log east rubble	1597+p	1740	inc	Unknown	
	NLC-1355	PNN	Loose log east rubble	1538+p	1737+	inc	Unknown	
	NLC-1356	PNN	Loose log east rubble	1611p	1734+	inc	Unknown	
	NLC-1357	PNN	Loose log east rubble	1632+p	1720	comp	Unknown	
	NLC-1358	POP	Loose log east rubble	NO DATE	—		Unknown	
	NLC-1359	PNN	Loose log east rubble	1609+p	1734	inc	Unknown	
	NLC-1360	POP	Loose log east rubble	NO DATE	—		Unknown	
	NLC-1361	PNN	Loose log east rubble	1586+p	1736	inc	Unknown	
	NLC-1362	PNN	Loose log east rubble	NO DATE	—		Unknown	

	NLC-1363	PNN	Loose log east rubble	1626p	1738	inc	Unknown	
	NLC-1364	PNN	Loose log west rubble	1638p	1728+		Unknown	
	NLC-1365	PNN	Loose log west rubble	1623+p	1737++	inc	Unknown	
GOMEZ	DNT-237	JUN	Roof primary RM 1	1617	1730+B	comp	MA ends, limb	
CANYON	DNT-238	JUN	Roof primary RM 1	NO DATE	—		MA ends, broken limb	SB
PUEBLITO	DNT-239	PNN	Roof primary RM 1	1610p	1734LG	inc	MA ends, limb, shape	SB; flat
LA 55831	DNT-240	JUN	Roof primary RM 1	1650	1731vv	inc	MA ends, limb	SB
	DNT-241	JUN	Roof primary RM 1	NO DATE	—		MA ends, limb	
	DNT-242	JUN	Roof primary RM 1	1608	1733L	inc	MA ends, bark, limb	
	DNT-243	JUN	Roof primary RM 1	1594	1735+L	inc	MA ends, limb, shape	flat
	DNT-244	JUN	Roof primary RM 1	1516+/-	1725+vv	inc	MA ends, limb, shape	flat
	DNT-245	JUN	Roof primary RM 1	1622p	1736G	inc	MA ends, shape	flat
	DNT-246	PNN	Roof secondary RM 3	1654p	1734L	inc	MA limb	
	DNT-247	PNN	Roof secondary RM 3	1638p	1734L	inc	MA limb	
	DNT-248	JUN	Roof secondary RM 3	NO DATE	—			split
	DNT-249	PNN	Roof secondary RM 3	1707	1735+vv	inc	MA limb	
	DNT-250	PNN	Hogan 1 or 2 Loose	1600	1710vv	inc	MA limb	
	DNT-251	PNN	Hogan 1 or 2 Loose	1680p	1733vv	inc	MA end, limb	
	DNT-252	PNN	Hogan 1 or 2 Loose	1404	1681+vv	inc	MA ends	
	DNT-253	PNN	Hogan 1 or 2 Loose	1509	1709vv	inc	MA ends, bark, imb	
	DNT-254	PNN	Hogan 1 or 2 Loose	1445	1725+vv	inc	MA limb, broke end	
	DNT-255	JUN	Hogan 1 or 2 Loose	NO DATE	—		MA shape, broke end	flat
	DNT-256	PNN	Roof primary (?) Fall? RM 3	1616p	1731+vv	inc	MA end, limb	
	DNT-257	PNN	Roof primary (?) Fall? RM 3	1639	1732vv	inc	MA end, limb	
	DNT-258	PNN	Roof primary (?) Fall? RM 3	1622p	1727vv	inc	MA ends, limb	
	DNT-259	PNN	Roof primary (?) Fall? RM 3	1507	1725vv	inc	MA end, limb, shape	flat
	DNT-260	PNN	Roof primary (?) Fall? RM 3	1416	1637+vv	inc	MA ends, limb	
	DNT-261	PNN	Roof primary (?) Fall? RM 3	1655p	1747vv	inc	MA ends, bark, limb	
	DNT-262	PNN	Roof primary (?) Fall? RM 3	1514	1706vv	inc	MA end, limb, shape	flat
	DNT-263	JUN	Hogan 3 Loose	1557	1726+vv	inc	MA end	
	DNT-264	PNN	Hogan 3 Loose	1883p	1938+GB		MA end	
	DNT-265	JUN	Hogan 4 Loose	1644	1723vv	inc	MA end, limb	
	DNT-266	JUN	Hogan 4 Loose	NO DATE	—		MA end	
	DNT-267	PNN	Loose log	1622p	1745vv	inc	MA end, limb	

Site Name Number	Sample Number	Species	Provenience	Inside Date	Outside Date	Terminal Ring	Tool Marks	Comment
HADLOCK'S	DNT-65	JUN	Upright post RM 3	1509	1655vv	inc		
CROW	DNT-66	PNN	Upright post RM 3	1551	1661vv	inc		
CANYON	DNT-67	JUN	Upright post RM 3	1494	1646vv	inc	MA limb	
LA 55830	DNT-68	JUN	Upright post RM 3	NO DATE			MA limb	
HOODED	DNT-51	PNN	Shelf pole RM 1	1612	1723G	inc	MA limb	SB
FIREPLACE	DNT-52	POP	Roof primary RM 1	NO DATE	—		MA limb	
LA 5662	DNT-53	PNN	Roof primary RM 2	1457	1722vv	inc	MA end, limb, shape	flat; SB
	DNT-54	PNN	Roof primary RM 2	1435	1721vv	inc	MA end, limb, shape	square; SB
	DNT-55	PNN	Roof primary RM 2	1501p	1722vv	inc	MA end, limb, bark	SB
	DNT-56	PNN	Roof primary RM 4	1388	1723vv	inc	MA end, limb, shape	flat; no SB
	DNT-57	PNN	Roof primary RM 3	1490	1723B	comp	MA limb, bark, shape	flat; no SB
	DNT-58	PNN	Roof primary RM 3	1398	1723B	comp	MA limb, bark	no SB
	DNT-59	PNN	Roof primary RM 3	1492	1723B	comp	MA limb, bark, shape	flat; no SB
	DNT-60	PNN	Roof primary RM 3	1440	1723B	comp	MA limb, bark, shape	flat; no SB
HUMMINGBIRD	DNT-740	PNN	RM 1 Roof primary	1692	1745v	inc	MA ends, limb	SB
HOUSE	DNT-741	PNN	RM 1 Roof primary	1619	1743+vv	inc	MA ends, limb	SB
LA 106575	DNT-742	DF	RM 1 Roof primary	1710p	1745v	inc	MA ends, limb	
	DNT-743	PNN	RM 1 Roof primary	1626	1743+v	inc	MA ends, limb	SB
	DNT-744	DF	RM 1 Roof primary	1674	1745c	inc	MA ends, limb	same as NLC-1295
	DNT-745	PNN	RM 1 Door lintel	1671	1745v	inc	MA ends	
	DNT-746	DF	RM 1 Door lintel	NO DATE	—		MA ends	
	DNT-747	DF	RM 1 Loose log	NO DATE	—		broken end	
	DNT-759	PNN	RM 1 Loose log	1643	1745vv	inc	MA limb, bark	
	DNT-748	PNN	RM 1 Loose log	1631p	1745vv	inc	MA ends, limb	
	DNT-749	PNN	RM 2 Loose log	1648	1744+vv	inc	MA ends, limb	
	DNT-750	PNN	RM 2 Loose log	1618	1740++vv	inc	MA ends, limb	same as NLC-1290
	DNT-751	PNN	RM 2 Loose log	1577+p	1741vv			same as NLC-1287
	DNT-752	PNN	RM 2 Loose log	—	—			same as NLC-1286
	DNT-753	PNN	RM 2 Loose log	1623	1742+vv	inc	MA limb	SB
	NLC-1279	PNN	Loose log outside RM 2	1618p	1730vv			same as DNT-756
	NLC-1280	PNN	RM 2 Loose log	—	—			same as DNT-755

	DNT-754	PNN	RM 3 Loose log	1607p	1731vv	inc		same as NLC-1282
	NLC-1281	PNN	RM 3 Loose log	NO DATE				
	NLC-1282	PNN	RM 3 Loose log	—	—			same as DNT-754
	NLC-1283	PNN	RM 3 Loose log	1638nc	1738	inc		
	DNT-755	PNN	RM 4 Roof primary?	—	—			same as NLC-1280
	DNT-756	PNN	RM 4 Loose log	—	—			same as NLC-1279
	NLC-1286	PNN	RM 4 Roof primary	—	—			same as DNT-752
	NLC-1287	PNN	RM 4 Roof primary	—	—			same as DNT-751
	NLC-1288	PNN	RM 4 Roof primary	1603+p	1742G	inc		
	NLC-1289	PNN	RM 4 Roof primary	1665+p	1745	inc		
	NLC-1290	PNN	RM 4 Roof primary	—	—			same as DNT-750
	NLC-1291	PNN	RM 4 Roof primary	1636p	1741G	inc		
	NLC-1292	PNN	RM 4 Loose log	NO DATE	—			
	NLC-1293	PNN	RM 4 Loose log	NO DATE	—			
	NLC-1284	PNN	RM 5 Door lintel	NO DATE	—			
	NLC-1294	PNN	RM 5 Roof primary	1573p	1744+G			
	NLC-1295	DF	RM 5 Roof primary	—	—			same as DNT-744
	NLC-1285	DF	RM 5 Roof secondary?	NO DATE	—			
	DNT-757	PNN	Forked Entryway	1543	1723vv	inc	MA limb, bark	
	DNT-758	PNN	Forked Entryway	1522	1723vv	inc	MA limb, bark	
	DNT-760	PNN	Loose Entryway fork	1635p	1723G	inc		same as NLC-1297
	DNT-761	PNN	Loose Entryway fork	1632	1741G	inc		same as NLC-1296
	DNT-762	DF	Entryway Loose log	1673	1744vv	inc	Limb broken	
	NLC-1296	PNN	Loose log	—	—	inc		same as DNT-761
	NLC-1297	PNN	Loose log	—	—			same as DNT-760
	NLC-1299	DF	Loose log	1661p	1743c			
	NLC-1298	PNN	Climbing pole?	1619p	1725G	inc		
JARMILLO	DNT-226	PNN	West door lintel	1686	1731+G	inc	MA ends, limb, bark	SB
CANYON	DNT-227	PNN	Middle door lintel	1679p	1730+G	inc	MA ends, limb, bark	SB
RUIN	DNT-228	PNN	East door lintel	1648p	1725vv	inc	MA limb, bark	no SB
KIN TSÉZÉÍ	DNT-853	PNN	loose beam		1687+vv	inc		weathered
LA 11092	DNT-854	JUN	loose beam		1643vv	inc	MA end	weathered
	DNT-855	JUN	loose beam		1661vv	inc	MA end	weathered
	DNT-856	JUN	loose beam	NO DATE	—		MA limb, ends	weathered

Site Name Number	Sample Number	Species	Provenience	Inside Date	Outside Date	Terminal Ring	Tool Marks	Comment
	DNT-857	JUN	loose beam	NO DATE	—		MA limb, end	weathered
	DNT-858	JUN	Ax-cut stump		1645vv	inc	MA cut	weathered
	DNT-859	JUN	Ax-cut limb	NO DATE	—		MA cut	weathered
KIN YAZHI	NLC-1248	PNN	Fireplace support	1601+p	1731	inc	Unknown	noncutting
LA 2433	NLC-1250	PNN	Interior loose log	1552+p	1742+		Unknown	noncutting
	NLC-1249	PNN	Loose log south wall	1515+p	1737+	inc	Unknown	noncutting
	NLC-1251	PNN	Loose wood fragment	1570+p	1734	inc	Unknown	noncutting
	NLC-1252	PNN	Loose wood fragment	1504+p	1730	inc	Unknown	noncutting
	NLC-1253	PNN	Loose log west side	1634+p	1741c		Unknown	
	NLC-1254	PNN	Loose log west side	1553+p	1730	inc	Unknown	noncutting
	NLC-1255	PNN	Loose log west side	1617p	1721	inc	Unknown	noncutting
	NLC-1256	PNN	Loose log west side	1646+p	1732	inc	Unknown	noncutting
	NLC-1257	PNN	Loose log west side	1606+p	1739cG		Unknown	
	NLC-1258	PNN	Scattered hogan wood	NO DATE	—		Unknown	
	NLC-1259	PNN	Scattered hogan wood	NO DATE	—		Unknown	
LA 2297	DNT-595	PNN	Loose log	1603	1714+G	inc	MA bark, limb	weathered
	DNT-596	PNN	Loose log	1608	1715G	inc	MA limb, end	weathered
	DNT-597	PNN	Loose log	1621	1724vv	inc	None	weathered
	DNT-598	PNN	Loose log	1593	1716vv	inc	None	weathered
	RG-818	PNN	Unknown	NO DATE	—			
	RG-819	PNN	Unknown	1541p	1716G	comp		
	RG-820	PNN	Unknown	1590p	1726vv	comp		
	RG-821	PNN	Unknown	1588p	1693vv	inc		
	RG-822	PNN	Unknown	1570p	1716vv	comp		
	RG-823	PNN	Unknown	1523p	1710vv	comp		
	RG-824	PNN	Unknown	NO DATE	—			
	RG-825	PNN	Unknown	NO DATE	—			
LA 2305	D-85	PNN	Unknown	1651np	1722vv	inc		
	D-220	PNN	Unknown	1675p	1719r	inc		
	D-221	PNN	Unknown	1612np	1718r	comp		

Site Name Number	Sample Number	Species	Provenience	Inside Date	Outside Date	Terminal Ring	Tool Marks	Comment
LA 8948	GOV-40	PNN	Ladder #1, North slope	1364p	1695vv	Inc	Unknown	
LOS PEÑOLES	GOV-41	PNN	Ladder #1, North slope	—	—			same as GOV-40
LARGO	NLC-1308	PNN	RM 2 Door lintel	1677+/-p	1737incG			
SCHOOL	DNT-61	POP	RM 2 Roof primary	NO DATE	—		MA ends, limb, bark	
LA 5657	NLC-1306	PNN	RM 2 Roof primary	NO DATE	—			
	NLC-1312	POP	RM 2 Roof (primary?)	NO DATE	—			
	NLC-1313	POP	RM 2 Roof (primary?)	NO DATE	—			
	NLC-1309	PNN	RM 2 Loose log	1617p	1736incG			
	NLC-1310	PNN	RM 2 Loose log	NO DATE	—			
	NLC-1311	POP	RM 2 Loose log	NO DATE	—			
	DNT-62	PNN	RM 3 Roof primary	1483p	1645vv	inc	MA ends, limb, bark, shape	squared
	NLC-1305	PNN	RM 3 Roof primary	1597p	1736c			
	DNT-63	POP	Loose log	NO DATE	—		MA ends	flat
	DNT-64	PNN	Loose log	1524	1731+vv	inc	MA ends,shape	
	NLC-1303	POP	Loose log	NO DATE	—			
	NLC-1304	PNN	Loose log	1550+p	1721+			
	NLC-1307	PNN	Loose log	NO DATE	—			
	NLC-1314	POP	Loose log	NO DATE	—			
MORRIS	GOV-49	JUN	Pueblito loose	1492	1661vv			
SITE 1	GOV-50	JUN	Pueblito loose	1549	1687+vv	inc		
LA 83,529	GOV-51	JUN	Pueblito loose	1420p	1726+vv	inc		
	GOV-52	JUN	Pueblito loose	1594p	1748+GB	inc		
	DNT-801	PNN	Loose log	1782	1901++vv	inc	MA end	
	DNT-802	PNN	Loose log	1661	1741vv	inc	MA end	
	DNT-803	JUN	Ax-cut stump	1461+/-	1747vv	inc	MA end	
	DNT-804	JUN	Ax-cut stump	NO DATE	—		MA end	
	DNT-805	PNN	Ax-cut stump	1634p	1748+G	inc	MA end	
	GOV-53	JUN	Sweatlodge loose	1150p	1332+vv	inc		
	GOV-54	JUN	Sweatlodge loose	1442	1652+vv	inc		
	GOV-55	JUN	Sweatlodge loose	NO DATE	—			

Site Name Number	Sample Number	Species	Provenience	Inside Date	Outside Date	Terminal Ring	Tool Marks	Comment
MUÑOZ	RG-687	PNN	Unknown	1499fp	1744+v	inc		
CANYON	DNT-418	PNN	Loose area 1	1525	1666vv	inc	MA limb, shape, ends	flat; secondary seats
LA 1687	DNT-419	PNN	Loose area 1	1480	1718vv	inc	MA limb	
	DNT-420	JUN	Loose area 1	1653	1718+vv	inc	MA limb, ends	secondary seats, charred
	DNT-421	PNN	RM 3 Roof primary (?)	1479	1734vv	inc	MA ends	fork
	DNT-422	JUN	RM 3 Roof primary (?)	1653	1743v	inc		
PABLO	DNT-229	JUN	Loose log	NO DATE	—		MA limb, end	
SPRING	DNT-230	JUN	Loose log	NO DATE	—		MA limb	
	DNT-231	JUN	Loose log	NO DATE	—		MA limb	
	DNT-232	JUN	Loose log	NO DATE	—		MA limb; burned end	
POINTED	DNT-574	PNN	RM 2 Roof primary	1430	1711vv	inc	MA limb, shape, ends	flat; spiral
BUTTE	DNT-575	PNN	RM 1 Roof primary	NO DATE	—		MA limb, shape, end	flat; spiral
LA 10733	DNT-576	PNN	RM 1 Roof primary	1431	1713vv	inc	MA limb, shape, end	flat
	DNT-577	DF	Primary loose log	1580	1713vv	inc	MA limb	spiral; same as DNT-578
	DNT-578	DF	Primary loose log	—	—		MA limb, end	flat; same as DNT-577
	NLC-1329	PNN	Hogan 2 fork	1502+p	1720+			
	NLC-1330	PNN	Hogan 2 leaner	NO DATE	—			
	NLC-1331	PNN	Hogan 4 leaner	1612p	1683inc			
	NLC-1332	PNN	Hogan 4 fork	1458p	1702+			
	NLC-1333	PNN	Hogan 4 fork	NO DATE	—			
	NLC-1334	PNN	Hogan 3 fork	1613p	1709c			
	NLC-1335	PNN	Hogan 3 leaner	NO DATE	—			
	NLC-1336	PNN	Hogan 3 leaner	1655+p	1721inc			
	NLC-1337	PNN	Hogan 3 leaner	NO DATE	—			
	NLC-1338	PNN	Hogan 3 leaner	1618p	1748+			
	NLC-1339	PNN	Corral element	1719p	1881+G	inc		
	NLC-1340	PNN	Corral element	1657p	1877+vv	inc		
	NLC-1341	PNN	Corral element	1713+/-p	1888+G	inc		
	NLC-1342	PNN	Passageway element	NO DATE	—			
	NLC-1343	PNN	Loose log east rubble	NO DATE	—			
	NLC-1344	PNN	Loose log (hogan?)	1550+p	1720inc			
	NLC-1345	PNN	Loose log (hogan?)	1396+p	1721+			

Site Name Number	Sample Number	Species	Provenience	Inside Date	Outside Date	Terminal Ring	Tool Marks	Comment
PORK	DNT-610	PNN	RM 1 Door lintel	1623	1731+L	inc	MA limb, ends	
CHOP	DNT-611	PNN	RM 1 Door lintel	1646	1735+L	inc	MA limb, ends	
PASS	DNT-612	PNN	RM 1 Roof primary	1618	1737+vv	inc	Burned end	SB
LA 5661	DNT-613	PNN	RM 1 Roof primary	1632	1739+vv	inc	MA end, burned end	SB
	NLC-1426	PNN	RM 1 (?) N Door lintel (?)	1609+p	1745B	comp	Unknown	
	DNT-614	PNN	RM 2 Roof primary	1672	1737+vv	inc	MA bark, limb, ends	
	DNT-615	PNN	RM 2 Roof primary	1645	1722vv	inc	MA limb, ends	
	DNT-616	PNN	RM 2 Roof primary	1629	1726vv	inc	MA limb, shape, ends	flat; fork
	DNT-617	PNN	RM 2 Roof primary	1621	1728vv	inc	MA limb, end	
	DNT-618	PNN	RM 2 Roof primary	1631	1733vv	inc	MA limb, shape, end	flat; fork
	DNT-620	PNN	RM 2 Door lintel	1653	1742vv	inc	MA limb, ends	
	NLC-1432	PNN	RM 2 Door lintel	1662+p	1741	comp	Unknown	
	DNT-619	PNN	RM 2 loose log	1614	1733vv	inc	MA limb	
	NLC-1433	PNN	RM 2 loose log	1608+p	1742	inc	Unknown	
	NLC-1434	PNN	RM 2 loose log	1602nc	1737+		Unknown	
	NLC-1436	PNN	RM 2? loose log	1600+p	1732+		Unknown	
	NLC-1437	PNN	RM 2 loose log	1604+p	1738	inc	Unknown	
	NLC-1438	PNN	RM 2 loose log	1632+p	1739	inc	Unknown	
	NLC-1427	PNN	RM 3 Loose log	1650nc	1744	inc	Unknown	
	NLC-1428	PNN	RM 3 Loose log	1610p	1745	inc	Unknown	
	NLC-1429	PNN	RM 3 Loose log	1604p	1742G		Unknown	
	NLC-1430	PNN	RM 3 Loose log	1611+p	1747	inc	Unknown	
	NLC-1431	PNN	RM 3 Loose log	1410+p	1725	inc	Unknown	
	NLC-1424	PNN	Loose log in room fill	1642p	1742G	inc	Unknown	
	NLC-1425	PNN	Loose log in room fill	1646p	1734+		Unknown	
	DNT-621	PNN	Corral south side	1846	1914+vv	inc	MA ends	fork; historic
	DNT-622	PNN	Corral south side	1716+p	1847++vv	inc	MA limb	
	NLC-1435	PNN	E ash dump wood frag	1633p	1740+		Unknown	
PUEBLITO	DNT-233	JUN	RM 1 Loose log	NO DATE	—		None	
CANYON	DNT-234	JUN	RM 2 Loose log	LOST	—		MA end	
LA 1684	DNT-235	JUN	RM 2 Loose log	1621	1725vv	inc	None	

Site Name Number	Sample Number	Species	Provenience	Inside Date	Outside Date	Terminal Ring	Tool Marks	Comment
	DNT-236a	JUN	Loose log	1653	1733vv	inc	MA limb	
	DNT-236b	JUN	Loose log	—	—		MA limb	same as DNT-236a
	NLC-845	PNN	Roof beam	1612+/-p	1732	inc	Unknown	
	NLC-846	PNN	Roof beam	1622+/-p	1732	inc	Unknown	
	RG-691	PNN	Unknown	1624p	1735c		Unknown	
PUEBLITO	NLC-994	PNN	Unknown	NO DATE	—			
EAST	NLC-993	PNN	Unknown	1586p	1713inc	inc	Unknown	
LA 55834	NLC-994	PNN	Unknown	1512p	1679++vv	inc	Unknown	
RABBIT	DNT-901	JUN	Door sill?	NO DATE	—			
HOUSE	DNT-902	JUN	door lintel	NO DATE	—			
LA 99804	DNT-903	JUN	Loose log east side	NO DATE	—			
	DNT-904	JUN	Loose log east side	NO DATE	—			
	DNT-920	JUN	Roof primary	1670p	1751+GB	inc	MA limb, ends	
	DNT-921	JUN	Roof primary	1555	1741++GB	inc	MA limb	
	DNT-922	JUN	Roof primary	1507	1669++vv	inc	Unknown	
	DNT-923	JUN	Roof secondary	1655	1752+GB	inc	MA end	
	DNT-924	JUN	Roof secondary	1475	1680++vv	inc	MA end	same as DNT-926
	DNT-925	JUN	Roof secondary	1540	1655G	inc	MA end	
	DNT-926	JUN	Roof secondary	—	—		MA ends	same as DNT-924
	DNT-927	JUN	Loose log below boulder west	1475	1680++vv		MA ends	
RIDGE TOP	NMS-2	JUN	House 3 Compound	1600+/-p	1687vv	inc		
HOUSE	NMS-3	JUN	House 3 Compound	1477+/-p	1720vv	inc		
LA 6287	DNT-638	PNN	House 3 (?) Loose log	1655	1721vv	inc	MA bark, limb, broken end	
	DNT-639	PNN	House 3 (?) Loose log	1618	1713vv	inc	MA limb, shape, end	flat
	DNT-640	JUN	South Room Door lintel	NO DATE			MA shape	split
	DNT-641	JUN	North Plaza Loose log	1640	1720vv	inc	Burned end	split
	DNT-642	PNN	North Plaza Loose log	1638	1720vv	inc	MA limb	

Site Name Number	Sample Number	Species	Provenience	Inside Date	Outside Date	Terminal Ring	Tool Marks	Comment
SANTO	RG-790-1	PNN	Unknown (Loose)	1511p	1726r	inc	Unknown	
NIÑO	RG-790-2	PNN	Unknown (Loose)	1438p	1746r	inc	Unknown	
CANYON	RG-790-3	PNN	Unknown (Loose)	1606p	1750r	inc	Unknown	
LA 2137	RG-790-4	JUN	Unknown (Loose)	1587fp	1746v	inc	Unknown	
SIMON	DNT-1	PNN	Roof primary	1635	1754	+G inc	MA ends, limb	
CANYON	DNT-2	JUN	Roof primary	NO DATE	—		MA limb	
LA 5047	DNT-3	JUN	Roof primary	NO DATE	—		MA limb	
	DNT-4	JUN	Roof primary	1519	1691	G inc	MA limb	
	DNT-5	JUN	Roof primary	1581	1638	G inc	MA limb	
	DNT-6	PNN	Roof primary	1668	1754	+G inc	MA ends, limb	
	DNT-7	JUN	Closing Material	NO DATE	—		None	
SPLIT	DNT-92	PNN	Loose ladder area	1632	1727vv	inc	MA ends	
ROCK	DNT-93	PNN	Loose ladder area	1610	1725vv	inc	MA limb	
RUIN	DNT-94	PNN	Loose ladder area	1427	1679+vv	inc	MA limb	
LA 5664	DNT-95	PNN	Loose ladder area	1408	1708+vv	inc	Notched; MA end	
	DNT-96	PNN	Loose ladder area	NO DATE	—			
	DNT-97	PNN	Hogan 1 leaner	1532	1665vv	inc		
	DNT-99	PNN	Hogan 1 leaner	NO DATE	—			
	DNT-98	PNN	Hogan 3 leaner	1398	1570+vv	inc		
TOWER OF	DNT-573	PNN	Horizontal intramural	1675P	1722v	inc	MA limb, end	
THE STANDING	DNT-568	PNN	Horizontal intramural	1651	1720vv	inc	MA limb	
GOD	DNT-569	PNN	Horizontal intramural	1651	1718vv	inc	MA end	
LA 55839	DNT-570	PNN	Horizontal intramural	1634	1722vv	inc	MA limb, ends	
	DNT-571	DF	Horizontal intramural	NO DATE	—		MA shape, ends	flat
	DNT-572	PNN	Loose log	1563	1704+vv	inc	MA shape	flat; spiral
TSÉATAH	DNT-928	PNN	Loose log south side	1574	1727vv	inc	Weathered	
	DNT-929	JUN	Loose log south side	1558	1647+vv	inc	Weathered	
	DNT-930	JUN	Loose log south side	1465	1605vv	inc	Weathered	
	DNT-931	JUN	Loose log north side	NO DATE			Weathered	

Site Name Number	Sample Number	Species	Provenience	Inside Date	Outside Date	Terminal Ring	Tool Marks	Comment
TSO'BITO	DNT-454	PNN	RM 1/4 East Door lintel	NO DATE	—			
LA 61134	DNT-455	JUN	RM 1/4 Middle Door lintel	1608+p	1700+LB	comp	MA limb	
	DNT-456	JUN	RM 1 shelf pole	1624	1682+vv	inc	Broken ends	
	DNT-457	PNN	Loose log south side	1637	1726v	inc	MA end, broken end	
TRUBY'S	DNT-516	PNN	RM 1 Roof primary	1621	1750+vv	inc	MA end; burned	
TOWER	DNT-517	PNN	RM 1 Roof primary	NO DATE	—		MA ends; burned	
LA 2434	DNT-518	PNN	RM 1 Roof primary	1586	1731++vv	inc	Burned in half	
	DNT-519	PNN	RM 1 Roof primary	1593	1749vv	inc	MA limb; burned	
	DNT-520	PNN	RM 1 Roof primary	1564	1752L	inc	MA limb, ends; burned	
	DNT-521	PNN	RM 1 Roof primary	1549	1751+B	inc	MA ends; not burned	
	DNT-522	PNN	RM 1 Roof primary	1683	1752LB	inc	MA ends; not burned	
	DNT-523	PNN	RM 1 Door lintel	1681	1753L	inc	MA bark	
	DNT-524	PNN	RM 1 Door lintel	1660	1752L	inc	MA bark	
	NLC-1316	DF	RM 1 Floor fill	NO DATE	—		Unknown	
	NLC-1322	PNN	RM 1 Ladder?	1607p	1745	inc	Unknown	
	NLC-1323	DF	RM 1 Ladder?	NO DATE	—		Unknown	
	DNT-525	PNN	RM 2 FP hood?	1484p	1730+vv	inc	MA bark; burned	
	DNT-526	PNN	RM 2 Roof primary	1583	1749+vv	inc	burned in half	
	DNT-527	PNN	RM 2 Roof primary	1607	1748+vv	inc	burned in half	
	DNT-528	PNN	RM 2 Door jamb	1619	1746+vv	inc	MA bark, shape	flat
	DNT-529	PNN	RM 2 Door lintel	1689	1747vv	inc	MA limb	
	DNT-530	PNN	RM 2 Door lintel	1613p	1745+L		MA limb	
	DNT-531	JUN	RM 2 loose log	NO DATE	—		MA end	
	DNT-532	PNN	RM 2 loose log	1391	1709++vv	inc	Burned end	
	DNT-533	PNN	RM 2 loose log	1484	1720++vv	inc	MA shape, ends	flat
	DNT-534	PNN	RM 2 loose log	1634	1720vv	inc	MA limb, ends	
	NLC-1315	PNN	RM 2 Floor fill	1467p	1748+G		Unknown	
	NLC-1317	PNN	RM 2 Floor fill	1431p	1732+G		Unknown	
	NLC-1318	DF	RM 2 Floor fill	NO DATE	—		Unknown	
	NLC-1319	PNN	RM 2 Floor fill	1692p	1743G		Unknown	
	NLC-1325	PNN	RM 6 Roof primary	1599p	1721	inc	Unknown	

NLC-1326	DF	RM 5 Roof primary	NO DATE	—		Unknown
NLC-1320	PNN	Floor fill	NO DATE	—		Unknown
NLC-1321	PNN	Floor fill	1665p	1743G	inc	Unknown
NLC-1324	PNN	Loose log	1429+p	1730	inc	Unknown
NLC-1327	PNN	Loose log	1589+p	1744	inc	Unknown
NLC-1328	PNN	Loose log	1620p	1752	inc	Unknown
DNT-1059	PNN	Loose log	1539p	1752rB	inc	MA end

Key
NLC= Navajo Land Claim Collection
DNT= Dinetah Collection
RG= Rio Grande Collection
D= Douglass Collection
NMB= Douglass Collection
NAV= Douglass Collection
GOV= Douglass Collection
NMS= Douglass Collection
JUN= Juniper
PNN= Pinyon
DF= Douglas-fir
PP= Ponderosa pine
POP= Cottonwood
MA= metal ax
imb= limb trim
bark= debarking
comp= complete
inc= incomplete
SB= smoke blackening present

References Cited

Ababneh, L. N., R. H. Towner, M. M. Prasciunas, and K. T. Porter

2000 The Dendrochronology of Palluche Canyon, Dinétah. *Kiva* 66(2): 267–89.

Adams, S. E.

1982 Salvage Excavation: The Sand Dune Site. Ms. on file, Southwest Regional Office, National Park Service, Santa Fe, N.M.

Ahlstrom, R. V. N.

1985 *The Interpretation of Archaeological Tree-Ring Dates*. Ph.D. dissertation, University of Arizona, Tucson. University Microfilms, Ann Arbor, Mich.

Ahlstrom, R. V. N., C. R. Van West, and J. S. Dean

1995 Environmental and Chronological Factors in the Mesa Verde–Northern Rio Grande Migration. *Journal of Anthropological Archaeology* 14:125–42.

Amsden, C. A.

1932 Navajo Origins. *New Mexico Historical Review* 7:193–209.

Anthony, D. W.

1990 Migration in Archaeology: The Baby and the Bathwater. *American Anthropologist* 92(4):895–914.

Austin, M. (editor)

1972 SAAD AHAAH SINIL. Dual Language. Navajo Curriculum Center, Rough Rock Demonstration School, Chinle, Ariz.

Ayer, Mrs. E. E.

1916 *The Memorial of Fray Alonso de Benavides, 1630*. Privately Printed, Chicago.

Bailey, L. R.

1964 *The Long Walk, a History of the Navajo Wars, 1846–68*. Great West & Indian Series 26. Westernlore Press, Los Angeles.

Bailey, G. A., and R. G. Bailey
1986 *A History of the Navajos: The Reservation Years*. School of American Research Press, Santa Fe, N.M.

Bandelier, A. F.
1890–92 *Final Report of Investigations Among the Indians of the Southwestern United States*. 2 vols. Archaeological Institute of America, Cambridge, Mass.

Bannister, B.
1962 The Interpretation of Tree-Ring Dates. *American Antiquity* 27(4): 508–14.

Bannister, B., J. S. Dean, and E. A. M. Gell
1966 *Tree-Ring Dates from Arizona E, Chinle-de Chelly–Red Rock Area*. Laboratory of Tree-Ring Research, University of Arizona, Tucson.

Bannister, B., J. W. Hannah, and W. J. Robinson
1966 *Tree-Ring Dates from Arizona K, Puerco–Wide Ruin–Ganado Area*. Laboratory of Tree-Ring Research, University of Arizona, Tucson.

Bartlett, K.
1932 Why the Navajos Came to Arizona. *Museum Notes* 5(6):29–32.

Basso, K. H.
1996 *Wisdom Sits in Places: Landscape and Language Among the Western Apache*. University of New Mexico Press, Albuquerque.

Basso, K. (editor)
1971 *Apachean Culture, History, and Ethnology*. Anthropological Papers of the University of Arizona No. 21. Tucson.

Baxter, L., K. J. Baird, L. C. Pedicino, and K. Scotti
1997 The Dendrochronology of the Reconstruction of Kinishba. *Tree-Ring Bulletin* 54.

Begay, R. M., and A. Roberts
1996 The Early Navajo Occupation of the Grand Canyon Region. In *The Archaeology of Navajo Origins*, edited by R. H. Towner, pp. 197–212. University of Utah Press, Salt Lake City.

Binggeli, R., C. Muceus, and J. Oda
1998 Four Thousand Years at One Campsite: Bull Canyon Revisited. In *Diné Bíkéyah: Papers in Honor of David M. Brugge,* edited by M. S. Duran and D. T. Kirkpatrick, pp. 29–38. Archaeological Society of New Mexico No. 24. Albuquerque.

Blomberg, B.
1983 *Mobility and Sedentism: The Navajo of Black Mesa, Arizona*. Center for Archaeologicial Investigations Research Paper No. 32. Southern Illinois University, Carbondale.

Bolton, H. E.
1950 Pageant in the Wilderness: The Story of the Escalante Expedition to the Interior Basin, 1776; Including the Diary and Itinerary of Father Escalante. *Utah Historical Quarterly* 18:1–265.

Bradford, J. E.
1987 It'll All Come Out in the Wash: A Report on Test Excavations at the Sand Dune Site, Hubbell Trading Post National Historic Site, Ganado, Arizona. Ms. on file, Branch of Cultural Resources Management, Division of Anthropology, Southwest Cultural Resources Center, National Park Service, Santa Fe, N.M.

Brown, G. M.
1990 Old Wood and Early Navajos: A Chronometric Analysis of the Dinétah Phase. Paper presented at 5th annual Navajo Studies Conference, Shiprock, N.M.
1991 *Archaeological Data Recovery at San Juan Coal Company's La Plata Mine, San Juan County, New Mexico.* Technical Report No. 355, Mariah Associates, Albuquerque, N.M.
1993 *Archaeological Testing at Three Sites Near Romine Mesa.* Technical Report No. 809–02. Mariah Associates, Albuquerque, N.M.
1996 The Protohistoric Transition in the Northern San Juan Region. In *The Archaeology of Navajo Origins*, edited by R. H. Towner, pp. 47–70. University of Utah Press, Salt Lake City.

Brown, G. M., J. A. Evaskavich, R. G. Holloway, and C. D. Wilson
1992 *Archaeological Investigations at Grassy Canyon: A Gobernador Phase Navajo Habitation in the Navajo Reservoir District.* Technical Report No. 535, Mariah Associates, Albuquerque, N.M.

Brown, G. M., and P. M. Hancock
1992 The Dinétah Phase in the La Plata Valley. In *Cultural Diversity and Adaptation: The Archaic, Anasazi, and Navajo Occupation of the Upper San Juan Basin*, edited by L. S. Reed and P. F. Reed, pp. 69–90. Cultural Resources Series No. 9. New Mexico Bureau of Land Management, Santa Fe.

Brown, K. L., and M. E. Brown
1995 Cultural Behavior and Past Environments as Reflected in the Vertebrate Faunal Assemblages. In *A Chapter in Early Navajo History: Late Gobernador Phase Pueblito Sites of the Dinetah District*, by M. Marshall, pp. 145–200. Office of Contract Archeology, University of New Mexico Report No. 185–469B. Albuquerque.

Brugge, David M.
1963 *Navajo Pottery and Ethnohistory.* Navajoland Publications Series, vol.2. Navajo Tribal Museum, Window Rock, Ariz.
1966a Events in Navajo History (No. 15): Jemez Sheltered Navajos during the Carson Campaign. *Navajo Times*, June 16, p. 8, columns 1, 2, 3.
1966b Events in Navajo History (No. 14): Origin of the Máiidesgizhnii Clan. *Navajo Times,* June 9, p. 9.
1966c Events in Navajo History (No. 16): Origin of the Nakaidinés and Toyahedliinii Clans. *Navajo Times,* September 8, p. 29b.

1968a *Navajos in the Catholic Church Records of New Mexico, 1694–1875*. Research Reports No. 1, Research Section, Navajo Tribe, Window Rock, Ariz.

1968b Pueblo Influence on Navajo Architecture. *El Palacio* 75:14–20.

1969 Pueblo Factionalism and External Relations, *Ethnohistory* 16(2): 191–200.

1972a Navajo and Western Pueblo History. *The Smoke Signal*, No. 25. Tucson Corral of the Westerners, Tucson, Ariz.

1972b *The Navajo Exodus*. Archaeological Society of New Mexico, Supplement No. 4. Albuquerque.

1974 Traditional History of Wide Reeds. In *Wide Reed Ruin*, by James Mount, pp. 128–34. Ms. on file, Hubbell Trading Post National Historic Site, Ganado, Arizona, and Museum of Northern Arizona, Flagstaff.

1981a Comments on Athabaskans and Sumas. In *The Protohistoric Period in the North American Southwest, A.D. 1450–1700*, edited by D. R. Wilcox and W. B. Masse, pp. 291–320. Anthropological Research Papers, No. 24, G. A. Clark, general series editor. Arizona State University, Tempe.

1981b *Navajo Pottery and Ethnohistory*. Navajo Nation Papers in Anthropology No. 4. Navajo Nation Cultural Resource Management Program, Window Rock, Ariz.

1983 Navajo Prehistory and History to 1850. In *Southwest*, edited by Alfonso Ortiz, pp. 489–501. Handbook of North American Indians, vol. 10, W. Sturtevant, general editor. Smithsonian Institution, Washington, D.C.

1984 The Protohistoric Among Non-Pueblo Groups of the Southwest. In *Collected Papers in Honor of Harry L. Hadlock*, edited by N. L. Fox, pp. 169–75. Papers of the Archaeological Society of New Mexico, No. 9. Archaeological Society of New Mexico, Albuquerque.

1985 *Navajos in the Catholic Church Records of New Mexico, 1690 to 1875*. Navajo Community College Press, Tsaile, Ariz.

1986 *Tsegai: An Archaeological Ethnohistory of the Chaco Region*. USDI National Park Service, Washington, D.C.

1992 Discussion of Athabaskan Research. In *Current Research on the Late Prehistory and Early History of New Mexico*, edited by B. J. Vierra, pp. 337–52. Special Publication 1. New Mexico Archaeological Council, Albuquerque.

1993 Eighteenth-Century Fugitives from New Mexico Among the Navajos. In *Papers of the Third, Fourth, and Sixth Navajo Studies Conferences*, edited by A. Roberts, J. Smith and J. Piper, pp. 279–83. Navajo Nation Historic Preservation Department, Window Rock, Ariz.

1994 Post-Contact Demographic Trends in New Mexico. In *Artifacts,*

Shrines, and Pueblos: Papers in Honor of Gordon Page, edited by M. Duran and D. T. Kirkpatrick, pp. 49–61. Papers of the Archeological Society of New Mexico No. 20. Albuquerque.

1996 Navajo Archaeology: A Promising Past. In *The Archaeology of Navajo Origins*, edited by R. H. Towner, pp. 255–72. University of Utah Press, Salt Lake City.

n.d. The Influence of Defensive Requirements on Navajo Settlement. In *Coping with Diversity: Diachronic Studies of Navajo Responses to Environmental, Social, and Economic Variability*, edited by D. B. Oswald and J. S. Dean. Prescott College, Prescott, Ariz.

Brugge, D. M., and J. L. Correll

1959 Navajo Land Claims Project, site forms. Ms. on file, Navajo Nation, Window Rock, Ariz.

Bultnick, G.

1968–69 Contributions to the Study of the Preservation of Mud-Brick Structures. *Mesopotamia* 5(3–4):445–73.

Cameron, C. M.

1995 Migration and the Movement of Southwestern Peoples. *Journal of Anthropological Archaeology* 14:104–24.

Carlson, R. L.

1965 *Eighteenth Century Navajo Fortresses of the Gobernador District*. Museum Series in Anthropology, vol.10. University of Colorado, Boulder.

Carrillo, C. M.

1992 Where Were the Sheep: The Piedra Lumbre Phase Revisited. In *Current Research on the Late Prehistory and Early History of New Mexico*, edited by B. J. Vierra, pp. 323–26. Special Publication 1. New Mexico Archaeological Council, Albuquerque.

Carroll, C. H.

1979 *The Public Service Company of New Mexico Seboyeta Pumped Storage Archaeological Study*. Draft Ms. submitted to the Public Service Company of New Mexico, Albuquerque.

Carter, T. H., and R. Pagniero

1966 Notes of Mud-Brick Preservation. *Sumer* 22:65–76.

Copeland, J. M., and H. C. Rogers

1996 In the Shadow of the Holy People: Ceremonial Imagery in Dinétah. In *The Archaeology of Navajo Origins*, edited by R. H. Towner, pp. 213–30. University of Utah Press, Salt Lake City.

Cordell, L. S., and V. J. Yannie

1991 Ethnicity, Ethnogenesis, and the Individual: A Processural Approach Toward Dialogue. In *Processural and Postprocessural Archaeologies: Multiple Ways of Knowing the Past*, edited by R. W. Preucel, pp. 96–107. Center for Archaeological Investigations, Occasional Paper No. 10. Southern Illinois University, Carbondale.

Correll, J. L.
1979 *Through White Men's Eyes: A Contribution to Navajo History: A Chronological Record of the Navajo People from Earliest Times to the Treaty of June 1, 1868.* 6 vols. Navajo Heritage Center, Window Rock, Ariz.

Correll, J. L., and W. F. Kurtz
1957 Navajo Land Claims project site forms. Ms. on file, Navajo Nation, Window Rock, Ariz.

Cummings, B.
1940 *Kinishba: A Prehistoric Pueblo of the Great Pueblo Period.* Hohokam Museums Association and the University of Arizona, Tucson.

D'Arrigo, R. D., and G. C. Jacoby
1991 A 1000-Year Record of Winter Precipitation from Northwestern New Mexico, USA: A Reconstruction from Tree-Rings and its Relation to El Niño and the Southern Oscillation. *The Holocene* 1:95–101.

Dean, J. S.
1969 *A Chronological Analysis of Tsegi Phase Sites in Northeastern Arizona.* Papers of the Laboratory of Tree-Ring Research No. 3. Univeristy of Arizona, Tucson.

1978 Independent Dating in Archaeological Analysis. In *Advances in Archaeological Method and Theory*, vol. 1, edited by M. B. Schiffer, pp. 223-55. Academic Press, New York.

1988 Dendrochronology and Paleoenvironmental Reconstruction on the Colorado Plateaus. In *The Anasazi in a Changing Environment*, edited by G. Gumerman, pp. 25–44. Cambridge Univeristy Press, Cambridge.

1996 Behavioral Sources of Error in Archaeological Tree-Ring Dating: Navajo and Pueblo Wood Use. In *Tree-Rings, Environment, and Humanity: Proceedings of the International Conference*, Tucson, Ariz. 17–21 May 1994, edited by J. S. Dean, D. M. Meko, and T. S. Swetnam, pp. 497–503. Radiocarbon, Tucson, Ariz.

Dean, J. S., and W. J. Robinson
1977 *Dendroclimatic Variability in the American Southwest, A.D. 680–1970.* Laboratory of Tree-Ring Research, University of Arizona, Tucson.

Dean, J. S., and S. C. Russell
1978 Navajo Wood Use Behavior. Ms. on file, Laboratory of Tree-Ring Research, University of Arizona, Tucson.

Dean, J. S., and R. L. Warren
1983 Dendrochronology. In *The Architecture and Dendrochronology of Chetro Ketl*, edited by S. H. Lekson, pp.105–240. Reports of the Chaco Center No. 6. National Park Service, Albuquerque, N.M.

Dittert, A. E., Jr.

1958a Salvage Archaeology and the Navajo Project: A Progress Report. *El Palacio* 65(2):61–72.

1958b *Preliminary Archaeological Investigations in the Navajo Project Area of Northwestern New Mexico.* Museum of New Mexico Papers in Anthropology No. 1, Navajo Project Studies No. 1. Museum of New Mexico and School of American Research, Santa Fe.

Dittert, A. E., Jr., J. J. Hester, and F. W. Eddy

1961 *An Archaeological Survey of the Navajo Reservoir District of Northwestern New Mexico.* School of American Research and Museum of New Mexico Monograph No. 23. Santa Fe.

Dougherty, J. D.

1980 *Refugee Pueblos on the Santa Fe National Forest.* Cultural Resources Report No. 2. Santa Fe National Forest, Santa Fe, N.M.

Douglass, A. E.

1938 Southwestern Dated Ruins: V. *Tree-Ring Bulletin* 5(2):10–13.

Dykeman, D.

1998 Early Navajo Land Use and Economy–The Perfect Southwest Adaptation. Paper presented at the 64th Annual Meeting of the Society for American Archaeology, Chicago.

Dykeman, D., R. H. Towner, and J. Feathers

2002 Correspondence in Tree-Ring Dating and Thermoluminescence Dating: A Protohistoric Navajo Pilot Study. *American Antiquity* 67(1): 145–64.

Dykeman, D., and J. Wharton

1994 The Morris Site 1 Early Navajo Land Use Study: An Alternative Data Recovery Plan for Sites LA 11196, LA 83529, and LA 88766 in Williams Services Unit 29–7 Gas Gathering System, Rio Arriba County, New Mexico. Ms. on file, Navajo Nation Archaeology Department, Farmington, N.M.

1996 Preliminary Report on The Morris Site 1 Early Navajo Land Use Study and the Plan for Phase 3-Focused Data Recovery. Ms. on file, Navajo Nation Archaeology Department, Farmington, N.M.

Eddy, F. W.

1966 *Prehistory in the Navajo Reservoir District, Northwestern New Mexico. Parts I and II.* Museum of New Mexico Papers in Anthropology 15. Museum of New Mexico Press, Santa Fe.

Elliot, M. L.

1982 *Large Pueblo Sites near Jemez Springs, New Mexico.* Cultural Resources Report No. 3. Santa Fe National Forest, Sante Fe, N.M.

Enloe, J. G., W. C. Allan, P. S. Grigg, A. T. Smith, and S. Peckham

1973 An Archaeological Inventory and Evaluation of Some Prehistoric and Historic Sites in the Upper San Juan Drainage, New Mexico, with

Special Emphasis on Sites in the San Juan Planning Unit, Albuquerque District, Bureau of Land Management. Ms. on file, Bureau of Land Management, Farmington, N.M.

Espinosa, J. M.

1942 *Crusaders of the Rio Grande*. Institute of Jesuit History, Chicago.

Fall, P. L., J. A. McDonald, and P. C. Magers

1981 *The Canyon del Muerto Survey Project: Anasazi and Navajo Archaeology in Northeastern Arizona*. National Park Service, Western Archaeological Center, Publications in Anthropology No. 15. Tucson, Ariz.

Farmer, M. F.

1942 Navajo Archaeology of the Upper Blanco and Largo Canyons, Northern New Mexico. *American Antiquity* 8(1):65-79.

1947 Upper Largo Navajo, 1700–1775. *The Kiva* 12:15–24.

Farrell, M.

1995 Seasonality of Recently Excavated Dinétah Phase Sites. Paper presented at the 8th Navajo Studies Conference, Farmington, N.M.

Ferguson, T. J., and E. R. Hart

1985 *A Zuni Atlas*. University of Oklahoma Press, Norman.

Fetterman, J.

1996 Radiocarbon and Tree-Ring Dating at Early Navajo Sites: Examples from the Aztec Area. In *The Archaeology of Navajo Origins*, edited by R. H. Towner, pp. 71–82. University of Utah Press, Salt Lake City.

Forbes, J. D.

1960 *Apache, Navajo, and Spaniard*. University of Oklahoma Press, Norman.

Freeman, D.

1970 *Report on the Iban*. Athlone Press, London.

Fritts, H. C.

1991 *Reconstructing Large-Scale Climatic Patterns from Tree-Ring Data*. University of Arizona Press, Tucson.

Gibson, T.

1990 Raiding, Trading, and Tribal Autonomy in Insular Southeast Asia. In *The Anthropology of War*, edited by J. Haas, pp. 125–46. Cambridge University Press, Cambridge.

Gilpin, D.

1993a Historical Archaeology on the Navajo Indian Irrigation Project. In *DÁA'ÁK'EH NITSAA: An Overview of the Cultural Resources of the Navajo Indian Irrigation Project, Northwestern New Mexico*, by L. E. Vogler, K. Langenfeld, and D. Gilpin, pp. 191–281. Navajo Nation Papers in Anthropology 29. Navajo Nation Archaeology Department, Window Rock, Ariz.

1993b Historic Period Colonization of the Northern Chaco Plateau, Northwestern New Mexico. Paper presented at the 58th Annual Meeting of the Society for American Archaeology, St. Louis.

1996 The Early Navajo Occupation West of the Chuska Mountains. In *The Archaeology of Navajo Origins*, edited by R. H. Towner, pp. 171–96. University of Utah Press, Salt Lake City.

Goodwin, G.

1942 *Western Apache Raiding and Warfare*. University of Arizona Press, Tucson.

Graves, M.

1982 Anomalous Tree-Ring Dates and the Sequence of Room Construction at Canyon Creek Ruin, Arizona. *The Kiva* 47(3):107–31.

Graybill, D. A.

1989 The Reconstruction of Prehistoric Salt River Streamflow. In *Environment and Subsistence*, by D. A. Graybill, D. A. Gregory, F. L. Nials, S. K. Fish, R. E. Gasser, C. H. Miksicek, and C. R. Szuter, pp. 5–23. The 1982–1984 Excavation at Las Colinas, Vol. 5. Archaeological Series 162. Arizona State Museum, University of Arizona, Tucson.

Gregg, S.

1991 Indirect Food Production: Mutualism and the Archaeological Visibility of Food Production. In *Between Bands and States*, edited by S. Gregg, pp. 203–15. Occasional Paper No. 9. Center for Archaeological Investigations, Southern Illinois Universty, Carbondale.

Gregory, D. A.

1981 Western Apache Archaeology: Problems and Approaches. In *The Protohistoric Period in the American Southwest, A.D. 1450–1700*, edited by D. R. Wilcox and W. B. Masse. pp. 257–74. Arizona State University Anthropological Research Papers No. 24. Tempe.

Grissino-Mayer, H.

1996 A 2129-Year Reconstruction of Precipitation for Northwestern New Mexico, USA. In *Tree-Rings, Environment, and Humanity: Proceedings of the International Conference*, Tucson, Ariz., 17–21 May 1994, edited by J. S. Dean, D. M. Meko, and T. S. Swetnam, pp. 191–204. Radiocarbon, Tucson, Ariz.

Gunnerson, D. A.

1956 The Southern Athabascans: Their Arrival in the Southwest. *El Palacio* 63:346–65.

Gunnerson, J. H., and D. A. Gunnerson

1971 Apachean Culture: A Study in Unity and Diversity. In *Apachean Culture, History, and Ethnology*, edited by K. Basso and M. Opler, pp.7–27. Anthropological Papers of the University of Arizona No. 21. Tucson.

Habicht-Mauche, Judith A.

1992 Coronado's Querechos and Teyas in the Archaeological Record of the Texas Panhandle. *Plains Anthropologist* 37(140):247–59.

Hackett, C. W.

1937 *Historical Documents Relating to New Mexico, Nueva Vizcaya, and*

Approaches Thereto, to 1773. Carnegie Institution of Washington, Washington, D.C.

Hall, E. T., Jr.

1944 Recent Clues to Athabaskan Prehistory in the Southwest. *American Anthropologist* 46:98–105.

1951 Southwest Dated Ruins: VI. *Tree-Ring Bulletin* 17:26-28.

Hall, E. T. Jr., and W. S. Stallings

1941 Field notes on file, Laboratory of Tree-Ring Research. University of Arizona, Tucson.

Hammond, G. P., and A. Rey

1940 *Narratives of the Coronado Expedition*. Coronado Historical Series 2. University of New Mexico Press, Albuquerque.

1953 *Don Juan de Oñate, Colonizer of New Mexico, 1595–1628*. University of New Mexico Press, Albuquerque.

1966 *The Rediscovery of New Mexico 1580–1594*. Coronado Historical Series 3. University of New Mexico Press, Albuquerque.

Hancock, P. M.

1992 Evidence of the Dinétah Phase in the La Plata River Valley, San Juan County, New Mexico. In *Current Research on the Late Prehistory and Early History of Northern New Mexico*, edited by B. J. Vierra, pp. 287–97. Special Publication No. 1. New Mexico Archaeological Council, Albuquerque.

1997 Dendrochronological Dates from the Dinétah. Paper presented at the Protohistoric Conference, Albuquerque, N.M.

Hancock, P. M., and B. P. Johnson

1988 An Archaeological Survey of Seven Proposed Well Pads near Cedar Hill and on Delgadita Mesa, San Juan and Rio Arriba Counties, N.M. DCA Technical Report 1334. Division of Conservation Archaeology, Farmington, N.M.

Hancock, P. M., T. M. Kearns, R. A. Moore, M. A. Powers, A. C. Reed, L. Wheelbarger, and P. A. Whitten

1988 *Excavations in the Middle La Plata Valley for San Juan Coal Company*. Studies in Archaeology No. 6. Division of Conservation Archaeology, Farmington, N.M.

Hannah, J. W.

1965 Tree-Ring Dates from the Morris Sites, Gobernador District, New Mexico. In *Eighteenth Century Navajo Fortresses of the Gobernador District*, by R. L. Carlson, pp. 109–13. University of Colorado Series in Anthropology No. 10. Boulder.

Hantman, J. L.

1983 Social Networks and Stylistic Distributions in the Prehistoric Plateau Southwest. Ph.D. dissertation, Department of Anthropology, Arizona State University, Tempe. University Microfilms, Ann Arbor, Mich.

Haskell, J. L.
1975 *The Navajo in the Eighteenth Century: An Investigation Involving Anthropological Archaeology in the San Juan Basin, Northwestern New Mexico.* Ph.D. dissertation, Department of Anthropology, Washington State University, Pullman.
1987 *Southern Athapaskan Migration A.D. 200–1750.* Navajo Community College Press, Tsaile, Ariz.

Hendricks, R., and J. P. Wilson
1996 *The Navajos in 1705: Roque de Madrid's Campaign Journal.* University of New Mexico Press, Albuquerque.

Herr, S., and J. J. Clark
1996 Causes and Consequences of Historic Migrations in the Puebloan Southwest. Paper presented at the 61st Annual Meeting of the Society for American Archaeology, New Orleans, La.

Hester, J. J.
1962a *Early Navajo Migrations and Acculturation in the Southwest.* Museum of New Mexico Papers in Anthropology, vol.6. Museum of New Mexico Press, Santa Fe.
1962b An Ethnohistoric Reconstruction of Navajo Culture, 1582–1824. *El Palacio* 69(3):130–38.

Hester, J. J., and J. Shiner
1963 *Studies at Navajo Period Sites in the Navajo Reservoir District.* Museum of New Mexico Papers in Anthropology No. 9. Santa Fe.

Hill, W. W.
1940 Some Navajo Culture Changes during Two Centuries. *Smithsonian Miscellaneous Collections* 100:395–415.

Hodge, F. W.
1895 The Early Navajo and Apache. *American Anthropologist* 8:223–40.

Hodge, F. W., G. P. Hammond and A. Rey
1945 *Fray Alonso de Bénavides' Memorial of 1634.* University of New Mexico Press, Albuquerque.

Hoijer, H.
1938 The Southern Athapaskan Languages. *American Anthropologist* 40(1):75–87.
1971 The position of the Apachean languages in the Athapaskan stock. In *Apachean Culture, History, and Ethnology*, edited by K. H. Basso and M. E. Opler, pp. 3–6. Anthropological Papers of the University of Arizona No. 21. University of Arizona Press, Tucson.

Hogan, P.
1989 Dinétah: A Reevaluation of Pre-Revolt Navajo Occupation in Northwestern New Mexico. *Journal of Anthropological Research* 45(1): 53–56.
1991 Navajo-Pueblo Interaction During the Gobernador Phase: A

Reassessment of the Evidence. In *Rethinking Navajo Pueblitos*. Cultural Resources Series, vol.8. USDI New Mexico Bureau of Land Management, Albuquerque.

Hovezak, T. D., and L. Sesler

1995 Architecture of Dinétah. Paper presented at the Eighth Navajo Studies Conference, Farmington, N.M.

Huscher, B. H., and H. A. Huscher

1942 Athabaskan Migration via the Intermontane Region. *American Antiquity* 8:80–88.

Jacobson, L., S. Fosberg, and R. Brewley

1992 Navajo Defensive Systems in the Eighteenth Century. In *Cultural Diversity and Adaptation: The Archaic, Anasazi, and Navajo Occupation of the Upper San Juan Basin*, edited by L. S. Reed and P. F. Reed, pp. 105–36. Cultural Resources Series No. 9, Bureau of Land Management, New Mexico State Office, Santa Fe.

James, C. D., III

1976 *Historic Navajo Studies in Northeastern Arizona*. MNA Research Paper No. 1. Museum of Northern Arizona, Flagstaff.

Jett, S. C., and V. E. Spencer

1981 *Navajo Architecture: Forms, History, Distribution*. University of Arizona Press, Tucson.

John, E. A.

1975 *Storms Brewed in Other Men's Worlds: The Confrontation of Indians, Spanish, and French in the Southwest, 1540–1795*. University of Nebraska Press, Lincoln.

Jones, K. L.

1988 *Excavation of the Sand Dune Site (AZ K:6:11) at Hubbell Trading Post National Historic Site, Ganado, Arizona*. Southwest Archaeological Consultants, Santa Fe. Prepared for Southwest Regional Office, National Park Service, Santa Fe, N.M.

Kelley, K. B.

1993 The Complexity of Navajo Origins. In *Navajo Country—Diné Bikéyah*, by J. C. Winter, K. Ritts-Benally, and O. Tamir, pp. B-1 to B-3. Across the Colorado Plateau: Anthropological Studies for the Transwestern Pipeline Extension Project, Volume VIII. Office of Contract Archaeology, University of New Mexico, Albuquerque.

Kelley, K. B., and H. Francis

1994 *Navajo Sacred Places*. University of Indiana Press, Bloomfield.

Kemrer, M. F.

1974 *The Dynamics of Western Navajo Settlement: An Archaeological and Dendrochronological Analysis*. Ph.D. dissertation, University of Arizona, Tucson. University Microfilms International, Ann Arbor, Mich.

1978 Maximizing the Interpretive Potential of Archaeological Tree-Ring

Dates. In *Discovering Past Behavior: Experiments in the Archaeology of the American Southwest*, edited by P. Grebinger, pp. 29–53. Gordon and Breach, New York.

1992 An Appraisal of the Piedra Lumbre Phase in North Central New Mexico. In *History and Ethnohistory along the Rio Chama*, prepared by J. D. Schelberg and R. R. Kneebone, pp.66–108. U.S Army Corps of Engineers, Albuquerque District, Albuquerque, N.M.

Kessell, J. L.

1979 *Kiva, Cross, and Crown*. National Park Service, U.S. Government Printing Office, Washington, D.C.

1987 *Kiva, Cross, and Crown: The Pecos Indians and New Mexico 1540–1840*. University of New Mexico Press, Albuquerque.

Keur, D. L.

1941 *Big Bead Mesa: An Archaeological Study of Navajo Acculturation, 1745–1812*. Memoirs of the Society for American Archaeology 1. Menasha, Wisc.

1944 A Chapter in Navajo-Pueblo Relations. *American Antiquity* 10(1): 75–86.

Kidder, A. V.

1913 Some Undescribed Ruins of the Historic Period from the Upper San Juan. *American Journal of Archeology* 2(17):88–90.

1920 Ruins of the Historic Period in the Upper San Juan Valley, New Mexico. *American Anthropologist* 22:322-29.

Kluckhohn, C., and D. Leighton

1962 *The Navaho*. Doubleday & Co., Garden City, N.Y.

Kulishek, J.

1996 Settlement Patterns, Population, and Congregación on the 17th Century Jemez Plateau. Paper presented at the 61st Annual Meeting of the Society for American Archaeology, New Orleans, La.

Lightfoot, R. R.

1992 Architecture and Tree-Ring Dating at the Duckfoot Site in Southwestern Colorado. *Kiva* 57(3):213–36.

Lummis, C.

1900 Fray Zarate Salmerons's Relacion. *Land of Sunshine* 12:39–48, 104–13, 180–87.

Magers, P. C.

1976 Navajo Settlement in Canyon del Muerto. Unpublished Ph.D. dissertation, University of Arizona, Tucson. University Microfilms, Ann Arbor, Mich.

Malcolm, R. L.

1939 Archaeological Remains, Supposedly Navajo, from Chaco Canyon, New Mexico. *American Antiquity* 5(1):4–20.

Marshall, M. P.

1979 Ceramic Analysis. In *The Public Service Company of New Mexico*

Seboyeta Pumped Storage Archaeological Study by C. Carroll. Draft Ms. submitted to the Public Service Company of New Mexico, Albuquerque.

1985 *The Excavation of the Cortez CO_2 Pipeline Project Sites, 1982–1983*. Office of Contract Archaeology, University of New Mexico, Albuquerque.

1991 The Pueblito as a Site Complex: Archaeological Investigations in the Dinétah District. In *Rethinking Navajo Pueblitos*. Cultural Resources Series, vol.8. USDI New Mexico Bureau of Land Management, Albuquerque.

1995 *A Chapter in Early Navajo History: Late Gobernador Phase Pueblito Sites of the Dinetah District*. Office of Contract Archaeology, University of New Mexico Report No. 185–469B. Albuquerque.

Matson, D. S., and A. H. Schroeder

1957 Cordero's Description of the Apache—1796. *New Mexico Historical Review* 32(4):335–56.

Matthews, W.

1897 Navaho Legends. *Memoirs of the American Folklore Society* 5.

1994 *Navaho Legends*. University of Utah Press, Salt Lake City.

McDonald, J. A.

1976 *An Archaeological Assessment of Canyon de Chelly National Monument*. Western Archaeological Center, Publications in Anthropology 5. Tucson, Ariz.

McGuire, R., and M. B. Schiffer

1983 A Theory of Architectural Design. *Journal of Anthropological Archaeology* 2:277–303.

McNitt, Frank

1972 *Navajo Wars, Military Campaigns, Slave Raids, and Reprisals*. University of New Mexico Press, Albuquerque.

Mera, H. P.

1939 *Style Trends in Pueblo Pottery 1500–1840*. Laboratory of Anthropology, Museum of New Mexico, Santa Fe.

Milich, A. R.

1966 *Relaciones by Zarate Salmerón*. Horn & Wallace, Albuquerque, N.M.

Mindeleff, C.

1898 Navaho Houses. *Seventeenth Annual Report of the Bureau of American Ethnology for the Years 1895–1896*, Pt. 2, pp. 469–517. Smithsonian Institution, Washington, D.C.

Moore, J. H.

1994 Putting Anthropology back Together Again: The Ethnogenetic Critique of Cladistic Theory. *American Anthropologist* 96(4):925–48.

Moorey, P. R. S.

1994 *Ancient Mesopotamian Materials and Industries*. Clarendon Press, Oxford.

Nash, S.
1997 A Cutting Date Estimation Technique for Ponderosa Pine and Douglas Fir Wood Specimens. *American Antiquity* 62(2):260–72.

Opler, M.
1983 The Apache Culture Pattern and Its Origins. In *Southwest*, edited by A. Ortiz, pp. 368–92. Handbook of North American Indians, vol. 10. W. G. Sturtevant, general editor. Smithsonian Institution, Washington, D.C.

Oswald, D. B.
1993 *Navajo Space Use Under Conditions of Increasing Sedentism*. Ph.D. dissertation, University of New Mexico, Albuquerque. University Microfilms International, Ann Arbor, Mich.

Palmer, W. C.
1965 Meteorological Drought. U.S. Weather Bureau Research Paper 45, U.S. Department of Commerce, Washington, D.C.

Parks, J. A., J. S. Dean, and J. L. Betancourt
n.d. Tree-Rings, Drought, and Pueblo Abandonment of South-Central New Mexico in the Late 1670s. Ms. on file, Laboratory of Tree-Ring Research, University of Arizona, Tucson.

Perry, R. J.
1991 *Western Apache Heritage: People of the Mountain Corridor*. University of Texas Press, Austin.

Powers, M. A., and B. P. Johnson
1987 *Defensive Sites of Dinétah*. Cultural Resources Series, vol.2. USDI New Mexico Bureau of Land Management, Albuquerque.

Reed, A. D., and J. C. Horn
1990 Early Navajo Occupation of the American Southwest: Reexamination of the Dinétah Phase. *Kiva* 55:283-300.

Reed, E. K.
1941 Information on the Navaho in 1706. *American Anthropologist* 43(3): 485–87.

Reed, L. S.
1995 Exploring Variability Within Early Navajo Ceramic Assemblages. Paper presented at the Third Annual Fruitland Conference, San Juan College, Farmington, N.M.

Reed, L. S., and P. F. Reed
1992 The Protohistoric Navajo: Implications of Interaction, Exchange, and Alliance Formation with the Eastern and Western Pueblos. In *Cultural Diversity and Adaptation: The Archaic, Anasazi, and Navajo Occupation of the Upper San Juan Basin*, edited by L. S. Reed and P. F. Reed, pp. 91–104. Cultural Resources Series No. 9, New Mexico Bureau of Land Management, Santa Fe.

1996 Reexamining Gobernador Polychrome: Toward a New Understanding of the Early Navajo Chronological Sequence in Northwestern

New Mexico. In *The Archaeology of Navajo Origins*, edited by R. H. Towner, pp. 83–108. University of Utah Press, Salt Lake City.

Reeve, F. D.

1956 Early Navaho Geography. *New Mexico Historical Review* 31:290–309.

1957 Seventeenth Century Navaho-Spanish Relations. *New Mexico Historical Review* 32:36–52.

1958 Navaho-Spanish Wars, 1680–1720. *New Mexico Historical Review* 33:204–231.

1959 The Navaho-Spanish Peace: 1720's–1770's. *New Mexico Historical Review* 34:9–40.

1960 Navaho-Spanish Diplomacy, 1770–1790. *New Mexico Historical Review* 35:200–35.

Reher, C. (editor)

1977 *Settlement and Subsistence along the Lower Chaco River: The CGP Survey*. University of New Mexico Press, Albuquerque.

Reichard, G. A.

1928 Social Life of the Navajo Indians. *Columbia University Contributions to Anthropology* 7. New York.

Reid, J. J., and S. M. Whittlesey

1989 *Byron Cummings' Architectural Reconstruction of Kinishba: An Archival Analysis*. Report submitted to the National Park Service and on file in the Department of Anthropology, The University of Arizona, Tucson.

Reynolds, W. E., N. S. Cella, and E. J. Caballero

1984 *Archaeological Investigation in the Gallegos Canyon Area: Blocks IV and V of the N.I.I.P.* Chambers Consultants and Planners, Albuquerque, N.M.

Riley, C. L.

1954 A Survey of Navajo Archaeology. *University of Colorado Studies in Anthropology* 4:45–60.

Robinson, W. J., and R. H. Towner

1993 *A Directory of Tree-Ring Dated Native American Sites in the American Southwest*. Laboratory of Tree-Ring Research, University of Arizona, Tucson.

Robinson, W. J., B. G. Harrill, and R. L. Warren

1974 *Tree-Ring Dates from New Mexico B: Chaco-Gobernador Area*. Laboratory of Tree-Ring Research, University of Arizona, Tucson.

Rocek, T. R.

1985 *Correlates of Economic and Demographic Change: Navajo Adaptations on Northern Black Mesa, Arizona*. Ph.D. dissertation, University of Michigan, Ann Arbor. University Microfilms International, Ann Arbor, Mich.

1995 *Navajo Multi-Household Social Units*. University of Arizona Press, Tucson.

Roessel, R. A., Jr.
1983 *Dinétah*. Navajo Curriculum Center, Rough Rock, Ariz.

Rothschild, N. A., B. J. Mills, T. J. Ferguson, and S. Dublin
1993 Abandonment at Zuni Farming Villages. In *Abandonment of Settlements and Regions*, edited by C. M. Cameron and S. A. Tomka, pp. 123–37. Cambridge University Press, Cambridge.

Russell, S. C.
1983 *The Navajo History and Archaeology of East Central Black Mesa*. Navajo Nation Papers in Anthropology, No. 21. Window Rock, Ariz.

Russell, S. C., and J. S. Dean
1985 The Sheep and Goat Corral: A Key Structure in Navajo Site Analysis. *The Kiva* 51:3–18.

Salzer, M. W.
2000 Temperature Variability and the Northern Anasazi: Possible Implications for Regional Abandonment. *Kiva* 65(1):295–318.

Sando, J. S.
1979a The Pueblo Revolt. In *Southwest*, edited by A. Ortiz, pp. 194–97. Handbook of North American Indians, vol. 9. W. G. Sturtevant, general editor. Smithsonian Institution, Washington, D.C.
1979b Jemez Pueblo. In *Southwest*, edited by A. Ortiz, pp. 418–29. Handbook of North American Indians, vol. 9. W. G. Sturtevant, general editor. Smithsonian Institution, Washington, D.C.

Sapir, E.
1916 The Na-Dene Languages, a Preliminary Report. *American Anthropologist* 17:534–58.
1936 Internal Linguistic Evidence Suggestive of the Northern Origin of the Navajo. *American Anthropologist* 38:224–35.

Schaafsma, C. F.
1975 *Archaeological Survey and Excavations at Abiquiu Reservoir, Rio Arriba County, New Mexico*. School of American Research, Santa Fe, N.M.
1976 *Archaeological Survey of Maximum Pool and Navajo Excavations at Abiquiu Reservoir, Rio Arriba County, New Mexico*. School of American Research, Santa Fe, N.M.
1978 Archaeological Studies in the Abiquiu Reservoir District. *Discovery* 2:41–69. School of American Research, Santa Fe, N.M.
1979 *The Cerrito Site (AR-4), A Piedra Lumbre Phase Settlement at Abiquiu Reservoir*. School of American Research, Santa Fe, N.M.
1981 Early Apacheans in the Southwest: A Review. In *The Protohistoric Period in the North American Southwest, A.D. 1450–1700*, edited by D. R. Wilcox and W. B. Masse, pp. 291–320. Anthropological Research Papers No. 24, G. A. Clark, general series editor. Arizona State University, Tempe.

1992 A Review of the Documentary Evidence for a Seventeenth-Century Navajo Occupation in the Chama Valley. In *Current Research on the Late Prehistory and Early History of New Mexico*, edited by B. J. Vierra, pp. 313–21. New Mexico Archaeological Council, Special Publication 1. Albuquerque.

1996 Ethnic Identity and Protohistoric Sites in Northwestern New Mexico: Implications for Reconstructions of Navajo and Ute History. In *The Archaeology of Navajo Origins*, edited by R. H. Towner, pp. 19–46. University of Utah Press, Salt Lake City.

2002 *Apaches de Navajo: Seventeenth-Century Navajos in the Chama Valley of New Mexico*. University of Utah Press, Salt Lake city.

Schaafsma, P.

1963 *Rock Art in the Navajo Reservoir District*. Museum of New Mexico Papers in Anthropology No. 7. Santa Fe.

Schiffer, M. B.

1986 Radiocarbon Dates and the "Old Wood" Problem: The Case of the Hohokam Chronology. *Journal of Archaeological Science* 13:13–30.

Schlanger, S.

1985 Prehistoric Population Dynamics in the Dolores Area, Southwestern Colorado. Ph.D. dissertation, Department of Anthropology, Washington State University, Pullman. University Microfilms International, Ann Arbor, Mich.

1990 Artifact Assemblage Composition and Site Occupation Duration. In *Perspectives on Southwestern Prehistory*, edited by P. E. Minnis and C. L. Redman, pp. 103–21. Westview Press, Boulder, Colo.

Schoenwetter, J., and F. W. Eddy

1964 *Alluvial and Palynological Reconstruction of Environments, Navajo Reservoir District*. Museum of New Mexico Papers in Anthropology No. 13. Museum of New Mexico Press, Santa Fe.

Scholes, F.

1937 Troublous Times in New Mexico, 1659–1670. *New Mexico Historical Review* 12(2).

Schroeder, A. H.

1963 Navajo and Apache Relationships West of the Rio Grande. *El Palacio* 70(3):5–23.

1965 A Brief History of the Southern Utes. *Southwestern Lore* 30:53–78.

1974 *A Study of the Apache Indians, Parts I–III*. Garland Publishing, New York.

Secoy, F. R.

1992 *Changing Military Patterns of the Great Plains Indians*. Bison Books, University of Nebraska Press, Lincoln.

Sesler, L., and T. D. Hovezak

1996 Navajo Settlement Patterns within the Fruitland Project Area. Paper presented at the Fourth Annual Fruitland Conference, April 19, 1996. Farmington, N.M.

Sesler, L. M., T. D. Hovezak, and R. H. Wilshusen

2000 Cultural Landscape of the Dinétah: The Navajo Occupation of Frances Mesa. In *Frances Mesa Alternative Treatment Project*, compiled by R. H. Wilshusen, T. D. Hovezak, and L. M. Sesler, pp. 159–254, La Plata Archaeological Consultants Research Paper No. 3. Dolores, Colo.

Spielmann, K. A.

1982 *Inter-Societal Food Acquisition among Egalitarian Societies: An Ecological Study of Plains/Pueblo Interaction in the American Southwest.* Ph.D. dissertation, Department of Anthropology, University of Michigan, Ann Arbor. University Microfilms International, Ann Arbor, Mich.

Stallings, S.

1937 Tree-ring dates on file, Laboratory of Tree-Ring Research, University of Arizona, Tucson.

Stokes, M. A., and T. L. Smiley

1963 Tree-Ring Dates from the Navajo Land Claim I. The Northern Sector. *Tree-Ring Bulletin* 25:8–18.

1964 Tree-Ring Dates from the Navajo Land Claim II. The Western Sector. *Tree-Ring Bulletin* 26:13–27.

1966 Tree-Ring Dates from the Navajo Land Claim III. The Southern Sector. *Tree-Ring Bulletin* 27:2–11.

1969 Tree-Ring Dates from the Navajo Land Claim IV. The Eastern Sector. *Tree-Ring Bulletin* 29:2–15.

Stubbs, S.

1930 Survey of the Gobernador Region. *El Palacio* 29(2):75–79.

Swadesh, F. L.

1966 *Hispanic Americans of the Ute Frontier from the Chama Valley to the San Juan Basin, 1694–1960.* Research Report No. 4, Tri-Ethnic Research Project. University of Colorado, Boulder.

Thomas, A. B.

1932 *Forgotten Frontiers: A Study of the Spanish Indian Policy of Don Juan Bautista de Anza, Governor of New Mexico, 1777–1787.* University of Oklahoma Press, Norman.

1935 *After Coronado: Spanish Exploration Northeast of New Mexico, 1696–1727.* University of Oklahoma Press, Norman.

Towner, R. H.

1992 Dating the Dinétah Pueblitos: The Tree-Ring Data. In *Interpreting the Past: Research with Public Participation*, edited by L. Jacobson and J. Piper, pp. 55–72. Cultural Resources Series, vol.10, USDI New Mexico Bureau of Land Management, Albuquerque.

1994 Big Bead Mesa in Regional Perspective. Poster presented at the 59th Annual Meeting of the Society for American Archaeology. Anaheim, Calif.

1996 The Pueblito Phenomenon: A New Perspective on Post-Revolt Navajo Culture. In *The Archaeology of Navajo Origins*, edited by R. H. Towner, pp. 149–70. University of Utah Press, Salt Lake City.

1997 *The Dendrochronology of the Navajo Pueblitos of Dinétah*. Ph.D. dissertation, Department of Anthropoogy, University of Arizona, Tucson. University Microfilms International, Ann Arbor, Mich.

2002 Archaeological Dendrochronology in the Southwestern United States. *Evolutionary Anthropology* 11(2):68–84.

Towner, R. H., and J. S. Dean

1992 LA 2298: The Oldest Pueblito Revisited. *Kiva* 59:315–31.

1996 Questions and Problems in Pre–Fort Sumner Navajo Archaeology. In *The Archaeology of Navajo Origins*, edited by R. H. Towner, pp. 1–18. University of Utah Press, Salt Lake City.

Towner, R. H., and B. P. Johnson

1998 *The San Rafael Canyon Survey: Reconstructing Eighteenth Century Navajo Population Dynamics in the Dinétah using Archaeological and Dendrochronological Data*. Arizona State Museum Archaeological Research Series, No. 170. University of Arizona, Tucson.

Towner, R. H., D. Grow, J. Psaltis, and A. Falzone

2001 The Importance of Sample Context in Dendroarchaeological Interpretation: An Example from Northwestern New Mexico, USA. *Tree-ring Research* 57(1):75–87.

Towner, R. H., H. C. Rogers, and J. M. Copeland

2001 Changing Threats, Changing Responses: Tactical Aspects of the Navajo Pueblitos of Dinétah. In *The Archaeology of Ancient Tactical Sites*, edited by J. Welch and T. Bostwick, pp. 111–30. Arizona Archaeologist No. 32. Phoenix.

Towner, R. H., L. Sesler, and T. Hovezak

1998 Navajo Culturally Modified Trees in the Dinétah. In *Diné Bíkéyah: Papers in Honor of David M. Brugge*, edited by M. Duran and D. Kirkpatrick, pp. 195–209. Archaeological Society of New Mexico, Albuquerque.

Twitchell, R. E.

1911 *The Leading Facts of New Mexican History*. 5 vols. The Torch Press, Cedar Rapids, Iowa.

1914 *The Spanish Archives of New Mexico*, Vols. 1, 2. The Torch Press, Cedar Rapids, Iowa.

Van Valkenburgh, R. F.

1940 We Found the Ancient Tower of Haskhek'izh. *Desert Magazine* 3:23– 25.

1941 *Dine Bikeyah*. The Navajo Tribe, Window Rock, Ariz. Mimeographed.

1974 Navajo Sacred Places. In *American Indian Ethnohistory: Indians of the Southwest*. Garland Publishers, New York.

Van Valkenburgh, R. F., and R. Yazzie

1956 Navajo Land Claims project site forms. Ms. on file, Navajo Nation, Window Rock, Ariz.

Vierra, B. J. (editor)

1992 *Current Research on the Late Prehistory and Early History of New Mexico*. New Mexico Archaeological Council, Special Publication 1. Albuquerque.

Vivian, R. G.

1960 The Navajo Archaeology of Chacra Mesa, New Mexico. M.A. thesis, Department of Anthropology, University of New Mexico, Albuquerque.

Weltfish, G.

1965 *The Lost Universe*. Basic Books, New York.

Wharton, J. T., D. D. Dykeman, and P. F. Reed

1996 *Small Site Archaeology in the Upper San Jaun Basin: Investigations at Archaic, Anasazi, and Navajo Sites in Northwestern New Mexico*. NNAD Fruitland Data Recovery Series No. 1. Navajo Nation Papers in Anthropology 32. Window Rock, Ariz.

Wilcox, D. R.

1981 The Entry of the Athabaskans into the American Southwest: The Problem Today. In *The Protohistoric Period in the American Southwest, A.D. 1450–1700*, edited by D. R. Wilcox and W. B. Masse. pp. 213–56. Arizona State University Anthropological Research Papers No. 24. Tempe.

1988 Avonlea and Southern Athapaskan Migrations. In *Avonlea Yesterday and Today*, edited by L. B. Davis, pp. 273–80. Saskatchewan Archaeological Society, Saskatoon.

Williams, A. W., Jr.

1970 *Navajo Political Process*. Contributions to Anthropology, vol. 9. Smithsonian Institution Press, Washington, D.C.

Wilshusen, R. H., T. D. Hovezak, and L. M. Sesler

2000 *Frances Mesa Alternative Treatment Project*. La Plata Archaeological Consultants, Research Papers No. 3, volume 1. La Plata Archaeological Consultants, Dolores, Colo.

Wilson, G. L.

1934 The Hidatsa Earthlodge. *Anthropological Papers of the American Museum of Natural History* 33(5):341–420.

Wilson, J. P.

1985 Before the Pueblo Revolt: Population Trends, Apache Relations and Pueblo Abandonments in Seventeenth Century New Mexico. In *Prehistory and History in the Southwest: Collected Papers in Honor of Alden C. Hayes*, edited by N. Fox, pp. 113–20. Papers of the Archaeological Society of New Mexico No. 11. Santa Fe.

Wilson, J. P., and A. H. Warren

1974 LA 2298 The Earliest Pueblito? *Awanyu* 2:8-26.

Winship, G. P.

1896 *The Coronado Expedition, 1540–1542*. Fourteenth Annual Report of the Bureau of Ethnology, 1892–1893, pp. 339–652. Washington, D.C.

Winter, J. C., K. Ritts-Benally, and O. Tamir

1993 *Navajo Country—Diné Bikéyah, Across the Colorado Plateau: Anthropological Studies for the Transwestern Pipeline Extension Project*, Volume VIII. Office of Contract Archeology, University of New Mexico, Albuquerque.

Winter, J. C., and P. Hogan

1992 The Dinétah Phase of Northwestern New Mexico: Settlement and Subsistence. In *Current Research on the Late Prehistory and Early History of New Mexico*, edited by B. J. Vierra, pp.299–312. New Mexico Archaeological Council, Special Publication 1. Albuquerque.

Worcester, D. E.

1947 *The Early History of the Navajo Indians*. Ph.D. dissertation, Department of History, University of California, Berkeley.

1951 The Navaho during the Spanish Regime in New Mexico. *New Mexico Historical Review* 26:101–18.

Wozniak, F. E.

1992 The Location of the Navajo Homeland in the Seventeenth Century: An Appraisal of the Spanish Colonial Records. In *Current Research on the Late Prehistory and Early History of New Mexico*, edited by B. J. Vierra, pp. 327–33. New Mexico Archaeological Council, Special Publication 1. Albuquerque.

Zolbrod, P. G.

1984 *Diné Bahane': The Navajo Creation Story*. University of New Mexico Press, Albuquerque.

Index

Note: Page entries printed in *italic* type refer to figures or tables.